BUILDING IN COB, PISÉ AND STABILIZED EARTH

C. Williams-Ellis

With an introduction by Gordon T. Pearson

Routledge
Taylor & Francis Group

LONDON AND NEW YORK

First published in 1916 by Cambridge University Press

Published by Donhead Publishing Ltd 1999

Published 2015 by Routledge
2 Park Square, Milton Park, Abingdon, Oxon OX14 4RN
711 Third Avenue, New York, NY 10017, USA

Routledge is an imprint of the Taylor & Francis Group, an informa business

© Taylor & Francis 1999
New introduction to this 1999 edition © Gordon T. Pearson

All rights reserved. No part of this book may be reprinted or reproduced or utilised in any form or by any electronic, mechanical, or other means, now known or hereafter invented, including photocopying and recording, or in any information storage or retrieval system, without permission in writing from the publishers.

Product or corporate names may be trademarks or registered trademarks, and are used only for identification and explanation without intent to infringe.

ISBN 13: 978-1-873394-39-7 (hbk)

A CIP catalogue for this book is available from the British Library

Donhead Publishing would like to acknowledge the help of Mike Chrimes at the Institute of Civil Engineers, London, in loaning an original copy of the work for this facsimile reprint.

BUILDING IN COB, PISÉ, AND STABILIZED EARTH

By

CLOUGH WILLIAMS-ELLIS, F.R.I.B.A.

and JOHN *and* ELIZABETH

EASTWICK-FIELD, A/A.R.I.B.A.

INTRODUCTION BY J. ST. LOE STRACHEY

LONDON: COUNTRY LIFE LIMITED

2-10 TAVISTOCK STREET, COVENT GARDEN, W.C.2

Pl. 1.—Cob House built by Ernest Gimson, Budleigh Salterton, Devon.

Acknowledgments

THE AUTHORS DESIRE TO ACKNOWLEDGE THEIR indebtedness to the Director, Building Research Station and to the Controller, His Majesty's Stationery Office, for permission to publish the text and illustrations of Chapter VIII; to Mr. A. Thorpe, for the drawings of his designs for Pisé cottages; Mr. F. Macdonald, for permission to use material and photographs from his booklet *Terracrete*; the Editor, *Journal of the Royal Institute of British Architects*, for permission to reproduce the illustrations and text on Russian practices; Mr. G. Gillanders and the Editor, *Journal of the Royal Sanitary Institute*, and Mr. A. F. Daldy, for permission to reproduce material and illustrations of Colonial practice; Mr. Ernst May, for material and illustrations on clay shingles; Mr. R. O. Mennell, for material and photographs of chalk building; Mr. B. H. Nixon and Miss J. Albery, A.R.I.B.A., A.M.T.P.I., of Hugh's Settlement, Quarley, for permission to reproduce photographs of cottages at Hugh's Settlement; Joseph Bradbury & Sons, Ltd., for the illustrations of G. E. Shuttering, and the Cement and Concrete Association for permission to reproduce photographs of soil-cement.

Also to Mr. K. P. R. Ehrenberg for help in translating from the German, and especially to Mr. A. W. Skempton of the Building Research Station and Mr. Hope Bagenal for advice and help.

Introduction to the 1999 Edition

Any student of British earthen buildings will soon become familiar with the name Sir Clough Williams-Ellis. In 1913 he planned to write a book on the results of his experiments with pisé. He had constructed a fruit house, a hospital extension and a wagon house, all to his satisfaction and considered that this method of construction was ideally suited to rural areas of Britain. World War I intervened and his writing was delayed. It was not until 1916 that he read the *Cyclopaedia or Universal Directory of Arts, Sciences and Literature* published by Abraham, Rees in 1819, that he realised that pisé was a common form of construction in the chalk areas of southern England. After the signing of the armistice in 1918, the need for his book was urgent and *Cottage Building in Cob, Pisé, Chalk and Clay* was published the following year.

The aftermath of the First World War left the United Kingdom short of one million homes. Coal, lime and cement were difficult to obtain and with a maximum possible annual output of only four billion bricks (i.e. one seventh of pre-war annual production), the need for an alternative form of construction was urgent. Sir Clough Williams-Ellis acknowledged that most post-war houses would be of brick but he considered that 'so far as rural housing is concerned the solution must be sought through the use of natural materials already existing on the site, materials that can be worked straight into the fabric of the building, without any elaborate or costly conversion, and that by local labour'.

At this time, each town had its own bye-laws to which all new building had to comply within its administrative area. Outside this area in the rural countryside, there were no restrictions. Hence the reason for Sir Clough Williams-Ellis's hypothesis.

His call for earthen building did not fall on deaf ears. His experiments with pisé had been widely reported in *The Spectator* magazine and had generated considerable interest, as the correspondence columns confirm. Lutyens and Gimson were designing buildings of earth and other books were to follow. The Department of Scientific and Agricultural Research published *Experimental Cottages: A Report on the Work of the Department at Amesbury, Wiltshire* by W. Jaggard in 1921, and *Building in Cob and Pisé de Terre* in 1922.

Following the Second World War, the United Kingdom found itself in the same economic situation as it had done after the First World War. The need to rebuild was urgent and Sir Clough Williams-Ellis decided to re-write his book, to expand on it in the light of recently acquired knowledge, to reflect upon his early experiments and to seek the assistance of two further authors. John and Elizabeth Eastwick-Field, both architects, collaborated with Sir Clough Williams-Ellis and *Building in Cob, Pisé and Stabilized Earth* was published in 1947.

The new edition, of which this is a facsimile, was increased from 125 to 164 pages. New chapters were added on adobe, stabilized earth, design, foreign practices and protective coverings.

Upon reading the book again, its relevance to the late twentieth century soon becomes apparent. Emphasis has changed from experimenting with earth construction to conserving the national earthen heritage, to which I made my contribution in 1992 with the publication of *Conservation of Clay and Chalk Buildings* (Donhead).

In the United Kingdom, there has been a reluctant response for calls to build with earth. The few examples of the last few years have concentrated on boundary walls, small extensions, bus and playground shelters, mainly in the West County. A few houses are planned, subject to compliance with the building regulations, but earth building has not been developed in the United Kingdom as it has in other countries. In Australia, the USA, Austria, France, Romania, India, Morocco and Peru, structures as diverse as conference centres, hotels, hospitals, sports centres, libraries, schools and airport terminals in addition to houses have been constructed. Some rise to five storeys and are constructed mainly of pisé de terre or adobe, sometimes in stabilized earth. In Switzerland, a Swiss Standard Specification has been produced for buildings of pisé construction.

In the preface, Sir Clough Williams-Ellis comments that 'it [pisé] has yet to prove itself in the fields of national housing and of competitive commercial building schemes on a large scale'. It is sad to think that eighty years after its publication, only now is the United Kingdom beginning to develop unburnt earth as a modern building material, many years behind its counterparts. The recent creation of a Centre for Earthen Architecture at the University of Plymouth has acted as a catalyst together with high profile conferences and publicity by groups such as ICOMOS (International Council on Monuments and Sites) and United

Kingdom government agencies. The rapid rise in the 'green' or sustainable architecture movement has also helped to stimulate interest in constructing new earthen buildings and it is to be hoped that this interest will be maintained and developed to encourage this country to return to its architectural roots.

Gordon T. Pearson
Stockbridge, Hampshire, June 1999

Contents

List of Illustrations

LIST OF ILLUSTRATIONS

Preface to Third Edition

THAT the demand for this book should still be so insistent as to call forth a third edition after the second had so long been out of print is clearly attributable to history repeating itself with an accuracy which is as singular as it is embarrassing.

Indeed, so similar are building conditions in this spring of 1947 to those obtaining after the first world war twenty-seven years ago, that the introduction then written by my late father-in-law (St. Loe Strachey of the *Spectator*) has been retained exactly as he wrote it as being still entirely and disturbingly appropriate. But the body of the book itself had become somewhat out of date through the extension of knowledge since its compilation, both through special laboratory research and the practice and study of actual building work in the field. Indeed, I was so well aware —in a general way—of the recent advances made, that I felt that a re-issue of the original book now would justly lay me open to a charge of idleness—of negligence amounting to fraud.

Not until I was assured of the co-operation of Mr. and Mrs. Eastwick-Field, with their special knowledge of current experimental building research, did I agree even to this heavily revised edition—because they, and they alone, were able to do what I myself could not even have attempted, and that is to bring things completely and reliably up to date.

Wherever the present book differs from or is better than the original, that is entirely due to them.

Of the very large number of letters that reached me from readers, quite ninety-nine out of every hundred were concerned with Pisé, and the interest displayed by Canada and Scandinavia particularly surprised me.

From both, I received many letters complaining of "the lumber shortage," and discussing the advantages of Pisé as compared with their traditional wood-construction.

If these great timber countries are themselves feeling the pinch, the advocates of wooden houses for England may find that they are not merely barking up the wrong tree, but up a tree that is not even there.

The timber famine is, in any case, a calamity to anyone

dependent on building, that is to everyone, for even a Pisé house must still have a roof and floors and joinery.

Pisé does at all events seem to offer us a more promising field for exploration than most of the other heterodox methods of construction that have been suggested, too often upon credentials that will not bear close scrutiny, though even now it is still probably only in its experimental infancy.

It has yet to prove itself in the fields of National Housing and of competitive commercial building schemes on a large scale.

Needless to say, Pisé does *not* claim to solve the housing problem. The best it can do, where conditions are suitable, is to substitute a natural, ready to hand and well-tried walling material for others of which, in any case, there is a desperate and almost universal shortage.

CLOUGH WILLIAMS-ELLIS.

April, 1947.

Introduction [1]

THE country is faced by a dilemma probably greater and more poignant than any with which it has hitherto had to deal. It needs, and needs at once, a million new houses, and it has not only utterly inadequate stores of material with which to build them, but has not even the plant by which that material can be rapidly created. There is not merely a shortage, but an actual famine everywhere as regards the things out of which houses are made. Bricks are wanted by the ten thousand million, but there are practically no bricks in sight. All that the brickyards of the United Kingdom can do, working all day and every day, is to turn out something like four thousand million a year. But to those who want houses at once, what is the use of a promise of bricks in five years' time? To tell them to turn to the stone quarries is a mere derision. Let alone the cost of work and of transport, it is only in a few favoured places that the rocks will give us what we want. Needless to say, we are short, too, of lime and cement, and probably shall be shorter. *No coal, no quicklime*, and *No coal, no cement*, and as things look now, it is going to be a case, if not of no coal, at any rate of much less coal. Even worse is the shortage in timber—the material hitherto deemed essential for the making of roofs, doors, windows, and floors. Raw timber is hardly obtainable, and seasoned timber does not exist. The same story has to be told about tiles, slates, corrugated iron, and every other form of "legitimate" roofing substance. There are none to be had.

In this dread predicament, what are we to do as a nation? What we must not do is, at any rate, quite clear. We must not lie down in the high road of civilization and cry out that we are ruined or betrayed, or that the world is too hard for us, and that we must give up the task of living in houses. Whether we like it or not, we have got to do something about the housing question, and we have got to do it at once, and there is an end. Translated into terms of action, this means that, as we have not got enough of the

---

[1] This introduction was written in 1919 for the first edition of this book by the then editor of *The Spectator*. There had been considerable correspondence in the columns of this paper on the subject of earth building.

As we know, the country was faced at that time with a similar shortage of building materials and of labour to that which faces us at present. (*J. & E. E.-F.*)

old forms of material, we must turn to others and learn how to house ourselves with materials such as we have not used before. Once again necessity must be the mother of invention, or, rather, of invention and revival, for in anything so old and universal as the housing problem it is too late to be ambitious. Here we always find that there has been an ancient Assyrian or Egyptian or a primitive man in front of us.

It is the object of the present book to attack part of the problem of how to build without bricks, and indeed without mortar, and, equally important, as far as possible without the vast cost of transporting the heavy material of the house from one quarter of England to another. That is my apology for introducing to the public a work dealing with what I can hear old-fashioned master-builders describing as the "bastard" forms of construction. One of these is Pisé de terre, the old system of building with walls formed of rammed or compressed earth: a system which was once known throughout Europe and of which the primitive tribesmen of Arizona and New Mexico knew the secret. Down to our own day it has been practised with wonderful success in the Valley of the Rhône. Then come our own Cob, once the cottage material *par excellence* of Devonshire and the West of England, our system of building with plain clay blocks, a plan indigenous in the eastern counties, and again the use of chalk and chalk Pisé.

PISÉ DE TERRE

FOR me Pisé de terre, ever since I heard of it, has offered special attractions. It, and it alone, provides, or, if one must be cautious, appears to provide, the way to turn an old dream of mine and of many other people into a reality. My connection with the problem of housing, and especially of rural housing, *i.e.* cottage housing, now nearly a quarter of a century old, has been on the side of cheap material. Rightly or wrongly (I know that many great experts in building matters think quite wrongly), I have had the simplicity to believe that if you are to get cheap housing, you must get it by the use of cheap material. It has always seemed to me that there is no other way. What more natural than first to ask why building material was so dear, and then what was the cause of its dearness? I found it in the fact that bricks are very expensive things to make, that stones are very expensive things to quarry, that cements are very expensive things to manufacture, and, worst of all, that all these things are very heavy and very expensive to drag about the country, and to "dump" on the site in some lonely situation where cottages or a small-holder's house and outbuildings are, to use the conventional phrase, "urgently demanded." Therefore, to the unfeigned amusement, nay, contempt of all my architectural friends, I spent a great deal of my leisure in the years before the war in racking my brains in the search for cheap material. My deep desire was to find something in the earth out of which walls could be made. My ideal was a man, or a group of men with spades and pickaxes, coming upon the land and creating the walls of a house out of what they found there. I wanted my house, my cottage in "Cloud-Cuckoo Land," to rise like the lark from the furrows. But I was at once dissuaded from my purpose by cautious and scientific persons. The chemists, if they did not scoff like the architects, were visibly perturbed. "Your dream is impossible," they said. "Nature abhors it as much as she used to be supposed to abhor a vacuum. If your soil is clay, and you can afford the time and cost of erecting kilns, and bringing coal to the spot to make the bricks, you can no doubt turn the earth on the spot into a house, but even then you had far better buy them of those that sell. Your dream of having some chemical which will mix with the earth and turn it into a kind of stone is the merest delusion. It is the nature of the earth to kill anything in the way of cement that is mixed with it. For example, even a little earth will kill concrete or mortar. Unless you wash your sand most carefully, and free it from all earth stain, you will ruin your concrete blocks." I appeared to be literally "up against" a brick wall. It was that or nothing. And then, and when things seemed at their very worst, a kind correspondent of *The Spectator* showed me a way of escape. I felt like a man lost in underground passages who suddenly sees a tiny square of light and knows that it means the way out. Somebody wrote, from South Africa I think, asking why I did not find the thing I wanted in Pisé de terre, much used in Australia, and occasionally in Cape Colony. Then came a rush of enlightenment. People who had seen and even lived in such houses wrote to *The Spectator*, and the world indeed for the moment seemed alive with Pisé de terre. I was even lent *The Farmer's Handbook* of New South Wales, in which the State Government provides settlers with an elaborate description of how to build in Pisé, and how to make the necessary shuttering for doing so. It was then, too, that I began to hear of the seventeenth- and eighteenth-century buildings of Pisé in the Rhône Valley. In fact, everybody but I seemed to know all there was to be known about Pisé de terre. For the moment, indeed, the situation seemed like that described in *Punch's* famous picture of the young lady and the German professor. *"What is Volapük?"* asks the young lady. *"Ze universal language,"* says the professor. *"Where is it spoken?"* *"No vairs."* Pisé de terre appeared to be the universal system of building, but as far as I could make out was practised "no vairs," or at any rate not in Europe.

I had got as far as the position described above, when down swept the war upon Europe, and everything had to be postponed in favour of the immediate need of filling the ranks of the nation's army and teaching the men how to fight our enemies. As the war went on, however, the demand for rapid, cheap, and temporary building became very great, and I felt I should be justified in trying some experiments with Pisé de terre, even in spite of the difficulty of obtaining labour.

I think I can best illustrate the nature of Pisé and what it can do, and I believe will do, if I shortly recount in chronological order these humble pioneer efforts.

In the summer of 1915 I found that it was necessary in the interests of the hospital established in my house to find a place in which to store apples, for the men in blue consumed them in incredible quantities. I thought I would try Pisé. Accordingly, I had some shuttering made on the Australian model—not splendid scientific shuttering such as is described in the body of this work, but still shuttering quite sufficient for the purpose. With great rapidity a little fruit-house was put up, roofed with boards, and covered with blocks of compressed peat in order to make a roof which would be both vermin-proof and also keep out the heat and the frost (Pl. 2). In my ignorance and my hurry, I now find that I violated every sound rule of Pisé construction. I built the walls during a week of rain, when the earth was wet, which was a great mistake; and I did not clear out the stones, which was another error that prevented the walls from being homogeneous. Worst of all, as soon as the walls were built (and very pretty walls they were, looking something like soft brown marble), I painted them over with tar, which, of course, would not enter the wet wall, but only made a skin, which in a few months peeled off exactly like the bark off a plane tree. Yet in spite of this ignorant mishandling of my material, the little fruit-house is still standing and sheltered till January the few apples Nature allowed us to gather last autumn. It looks disreputable, but there has been no structural collapse, nor will there be.[1]

No sooner was the fruit-house finished than I was met by the demands of my wife, the commandant of the hospital, to add to my house a patients' dining-room, which would be bright, dry, airy, warm, and comfortable, and be large enough for forty men to have their meals in, and to use as a sitting-room during the rest of the day. The local builder said that it was impossible to make a wooden addition, for there was no wood to be procured, or to build in bricks, since my house stands 600 ft. above the sea on an isolated chalk down. Crœsus would have found it difficult at that time to build on my site, and for the ordinary economic man —"L'homme à quarante écus"—it was quite impossible. But the room had got to be built, for the men were there, and built at once, since the out-of-door life of July and August could not continue. There was nothing to do but to fall back upon Pisé. I decided to be ambitious and to experiment, not merely in Pisé de terre, but what I then thought—and perhaps rightly—was a new form of Pisé, i.e. Pisé de craie, or compressed chalk. My shuttering, therefore, was put up. A hole not very far off was dug in the earth, the chalk which was almost at the surface was quarried out, and we began to build the wall, candid and contemptuous friends telling us, of course, that the chalk wall would never stand the frosts in so exposed a position, and that the wall, if made, would certainly explode! Everyone worked at that wall: the nursing staff, the coachman, an occasional visitor, a schoolboy, a couple of boy scouts, members of the National Reserve who were guarding a "vulnerable point" close by, and even some of the patients. Patients as a rule will endure any toil with the utmost good temper if it is for the purposes of sport. If the task is useful, it does not interest them. Still, a wall which might explode offered a certain attraction. We worked with more zeal than discretion, but happily I had it in my mind that homogeneity was the essential,

[1] It still looks "disreputable" and there has still been no structural collapse, although the roof has been replaced by one of corrugated iron. Its condition in 1945, after thirty years' exposure to the weather, may be judged by the accompanying photographs (Pls. 4 & 5). The Army used it as a store-house during the 1939–45 war. (J. & E. E.-F.)

and therefore the hard nuggets of chalk as they were thrown into the shuttering to be compressed by the rammers were first chopped up with spades, much as one minces meat. The wall had no foundations. In Pisé you can make your foundations, so to speak, as you go, through the simple process of ramming.[1] Anyway, and to cut a long story short, the wall was made, was able to receive the roof, for which happily the local builder found some material, and not only did the wall stand, but showed a very creditable exterior. Its weight was, of course, enormous, for there were some twenty tons of chalk put into it. In spite of the irregularity of the labour, it did not take more than ten or twelve days to build. To prevent the wet and frost getting into it, I painted the main front with a patent liquid material for rendering walls damp-proof. The Chalk Pisé wall not only served its purpose, but served it very well. The room proved extraordinarily warm and comfortable, largely owing, no doubt, to the fact of a solid, very dry, 18-in. wall on the north-east side.[2]

My next venture was in response to an urgent appeal from a farm tenant to build him a wagon-house. The result is seen in the accompanying illustration (Pl. 6). This building, about 40 ft. by 30 ft., was made purely of earth, but some experiments were tried in the way of introducing hurdles into the shuttering in order to afford a surface to which plaster could easily cling. Suffice it to say that the plain earth, without plaster or any covering, more than justified itself. One part of the wall is very much exposed to the weather, but it has stood the rains and the frosts of three very bad winters without turning a hair. Lovers of the picturesque may like to know that it presents a pleasant face of light ochre, upon which a pale green efflorescence of lichen has appeared of late. Anyway, the frost has not touched it.[3]

Next I made some experiments in chalk farmyard walls. Unfortunately, however, one of these, which was not made homogeneous by chalk mincing, i.e. in which the nuggets of chalk were

[1] In practice it is found that good foundations and damp-proof courses are essential in earth construction.

[2] This room was pulled down in 1926 and replaced by a larger brick-built room. It is said by an eye-witness that the walls were still sound at that time and proved difficult to demolish.

[3] The wagon-house still stands in 1945, although it was built with no foundations and has received no maintenance. For the most part the plaster still adheres to the hurdles, and even those parts of the walls which have had no protection are still in relatively good condition. The durability of these walls and those of the fruit-house is due most likely in no small part to the wide overhang to the eaves (Pls. 4 & 6). (*J. & E. E.-F.*)

not properly broken up, got the wet into it, and true to the candid friend's prophecy did literally explode in the big frost of 1917–18. Another very pretty chalk wall is, however, standing to this day. But though Chalk Pisé will, I think, do well if properly made and properly protected, it is somewhat of a doubtful material for anything except a building with a good overlay of roof. Another structure put up by me was a largish gardener's potting shed. This was built purely of earth, and in dry weather. When the walls were perfectly dry, the local road authorities kindly came with their tar spray and sprayed it with hot tar, with most excellent results. The hot tar really entered instead of making a skin, with the result that the external walls thus treated resembled a section of tarred road stood up on end.

I may add that I lent my Pisé shuttering to a Guildford Volunteer Battalion, who in a ten-hour day, or rather, two days of five hours each, built an excellent hut about 20 ft. square and 10 ft. high, and thus showed that a platoon might house themselves with Pisé in a day, provided they had roofing material ready. This building had subsequently to be destroyed, because the ground on which it stood was wanted for another purpose. When it was knocked down the house-breakers were astonished at the strength and tenacity of the walls. Yet the earth out of which they were made was particularly bad—as one of the volunteers expressed it, not earth, but merely leaf-mould and horse-manure. The site had, as a matter of fact, been a suburban garden for at least two hundred years.

Before I leave the record of these terrestrial adventures I may note that in the early stages I received a great deal of encouragement from General Sir Robert Scott-Moncrieff. He was, indeed, so much struck by them that he drew up a series of instructions for walls of Pisé work which were issued to all engineer companies at the front in case they might have opportunities for experimenting. These instructions were based upon the Australian book and embodied the very simple form of shuttering there recommended.

There is one thing more to be said about Pisé. I believe that a useful development of the system may be found in the plan of ramming earth into moulds and making earth blocks, something like concrete blocks. Moulds of this kind are easy to make and are specially suitable when the soil is somewhat clayey in its nature. They have the advantage of being much cheaper than shuttering, and of being capable of being handled by one man without assistance. With a strong wooden mould and a good rammer a small-holder may easily build his own pigsty, his own chicken-house, and all the small outbuildings he requires, if not indeed add an extra room to his house. I am at present experimenting with these blocks, and only yesterday had the pleasure of seeing a sergeant (R.A.M.C.), discharged through ill-health and now trying to turn himself into a small-holder, building a pigsty with the help of one of my moulds.

Apropos of the elusive universality and yet non-existence of Pisé work, the following personal anecdote or footnote to compressed earth may amuse my readers. Happening to be sleeping in a bedroom at Brooks's Club in 1916, I noticed a charming Regency bookcase full of old books. Among them was a copy of a *Cyclopædia* of 1819. I thought it would be amusing to see whether there was any mention of Pisé de terre. What was my astonishment to find that what I thought was my own special and peculiar hobby and discovery was treated therein at very great length and with very great ability, but treated not in the least as anything new or wonderful, but instead as "this well-known and greatly appreciated system of building," etc., etc. To complete the irony of the situation, the fact was mentioned that a Mr. Holland had lately sent to the Board of Agriculture a memorandum as to how to put up houses and farm-buildings in this form of construction. My hair rose on my head, for I had just committed a similar official indiscretion myself, and had been bombarding appropriate authorities with what I thought must be a complete novelty. Truly one can never be first or do anything new. It is always "in the Files," as Mr. Kipling says. Even in our most original moments we only keep on feebly imitating somebody else. The claim to originality is nothing but a muddy mixture of pride and ignorance. What did, however, somewhat amaze me was the calm statement of the *Cyclopædia* that this system of building was now well known in the counties of—and then came the names of practically all the counties of Southern England. And yet I had been keenly on the look-out for such buildings for several years. The cynic will say that they had all fallen down. That only shows the weakness of the cynic's point of view. The truth is, they are often concealed under various disguises of plaster, paint, and weather tiles. Few people know what their own walls are

Pl. 2.—The beginning of a Pisé Fruit-house.

Pl. 3.—The Fruit-house completed with roof of peat blocks on
rough boarding.

Pl. 4.—The Fruit-house in 1945, thirty years after erection.

Pl. 5.—Close-up view of the Pisé walling.

Pl. 6.—The Newlands Wagon-house.   Interior.

Pl. 7.—The Wagon-house in 1945, showing the plastering on
hurdles fixed to the wall.

Pl. 8.—Wayside Station of Pisé at Simondium, South Africa, designed by the late Sir Herbert Baker.

really made of till they try to cut a new opening for a door or a window in them.

COB AND CHALK

OF Cob I know little by actual experiment. It is fully dealt with in the body of this work, and readers will find that it is a kind of mud or clay concrete reinforced with straw. It is therefore totally and absolutely different from Pisé. One is wet, the other dry.

All that need be said about chalk is said by the author of the present book.

A POSTSCRIPT

IN the body of this work, mention is made of a very successful experiment in Pisé de terre made by the officials of a Rhodesian mining company; the outcome, I am proud to think, of my pre-war advocacy of Pisé in *The Spectator*. No sooner had my introduction been finished than there came by way of postscript an exceedingly interesting series of photographs, sent to me by Mr. Pickstone, a gentleman very well known in South Africa for his fruit gardens, his peaches, and his apricots. On the strength of what he had read in *The Spectator*, Mr. Pickstone lately undertook to build a station building and station-master's house for the railway-station at Simondium in the Drakenstein Valley, a place which during the summer is noted for its great heat. In the January number of the *South African Railways and Harbours Magazine*, Mr. Pickstone gives a detailed account of his bold and successful experiment, and illustrates it by a reproduction of some of his photographs (Pl. 8). Here is his own account of what he did.

"It must have been about eighteen months ago that the railway administration decided to promote Simondium Siding to the dignity of a station. As a siding, it had always been a busy place in the fruit season, during which time a permanent checker had for some years been kept quite busy, his accommodation being a couple of small tin shanties, and he had been accustomed to board out where he could. Now we were to have a 'pukka' station-master and, presumably, suitable premises. The department quickly got to work and the station-master's house arrived. It was what one might call a second-hand, or even a third- or fourth-hand one, consisting of the inevitable sheets of galvanised iron and the ever-essential Oregon and Swedish timber. Our new station-master also shortly afterwards arrived, and turned out to be a married man with a wife and four children. The station-master was not a grouser, but during the hot summer—and it is terribly hot in the Drakenstein Valley during that time of year—he complained to me that it was almost impossible to hold on, owing to the conditions under which he and his family had to live. It was just about this time that I saw in *The Spectator* a series of articles strongly advocating 'Pisé de terre' construction for buildings of all kinds; especially was it recommended as a war-time expedient for rapid and economical construction for barracks and hospitals, and, indeed, it was strongly recommended by Mr. St. Loe Strachey, the editor, for all sorts of general building and military purposes. It is a curious fact, which many readers could verify, that frequently one lives one's life under certain conditions, and in reality remains absolutely blind to their presence and potentialities. Here was I, living in a country where some of the most beautiful old homesteads are on the principle of the 'Pisé de terre' construction, and a large proportion of the older farm buildings in this district also built of similar material, with the additional pleasing accompaniment of beautiful beams, ceilings, and floors made of colonial pine—one may advisedly add, the *despised* colonial pine. Some of these buildings have stood the wear and use of close on a century, and are still an object of joy to those privileged to have an eye to see. Here lived I, as I say, blind to its potentialities for today, although it had been clearly appreciated and carried out with the most charming and solid results by our great-grandfathers in the old slave-labour days."

The supervising architect, Mr. Kendall, who was responsible for carrying out the work to the admirable design of Mr. Herbert Baker, gives the following description of the way the work was actually executed, which contains several very useful hints:

"The construction of walls determined upon was that known as 'Pisé de terre,' consisting of earth walls some 18 in. to 24 in. thick, which owe their solidity to a simple process of ramming between wooden casings previously placed in position on both sides. These walls are built in stages of some 3 ft. in height, the wood casing being raised at intervals as required. The frames for doors and windows are placed in position at the right time, and anchored into the walls by means of long hoop iron ties. These walls, when completed, give a surface almost as hard as burnt brick, but the external angles present a slight point of weakness, as from their exposure they would be naturally inclined to chip away in cases of rough usage. In order to overcome this it was arranged that irregular brick quoins should be embedded in the angles all the way up as the work proceeded. The walls, when completed, were then plastered and whitewashed, and present as good an appearance as more expensively plastered brickwork. As additional security the weather sides were given, prior to whitewashing, a coat of hot gas tar direct on the plaster, which in all exterior work was lime plus 10 per cent. cement. The roofs are of thatch with a fairly good overhang at the eaves in order to form a protection for the walls."

On one point, however, we may reassure Mr. Kendall. I do not think he need be afraid of his walls being destroyed by the weather even if he has no overhang. Part of a Pisé wall in my cart-shed, built in a very exposed situation, has no overhang. Further, the wall is not covered by cement or any other protective covering. The compressed earth was left quite bare, and yet the three worst winters of alternating wet and frost have made no impression upon the wall. It seems to be both rain-proof and frost-proof.[1]

I may add that Mr. Pickstone informs me in a letter dated February 19th that the Pisé walls have proved an enormous success from the point of view of protection from the heat. Whereas in an iron building lined with wood the temperature in the hot weather went up to 104 degrees Fahrenheit, in the station-master's Pisé de terre dining-room the thermometer registered only 86 degrees. Those who have ever lived where such temperatures prevail will note the immense advantage gained by the Pisé walls. Such temperatures try strong men and women, and for children they are positively death-dealing. With so successful an experiment as that at Simondium before my eyes, I am beginning to feel that I may live to correct my views that this universal system of building is practised "no vairs."

PLINY ON PISÉ DE TERRE

Now for something which I have kept as the *bonne bouche* of my earthy story. At the end of my researches and experiments I

---

[1] The provision of a generous overhang is, however, to be most strongly recommended. (*J. & E. E.-F.*)

COB, PISÉ, AND STABILIZED EARTH

found that Pliny has got it all in his *Natural History* in six lines!
There is no need for more words.

"*Have we not in Africa and in Spain walls of earth, known as
'formocean' walls? From the fact that they are moulded, rather than
built, by enclosing earth within a frame of boards, constructed on either
side. These walls will last for centuries, are proof against rain, wind, and
fire, and are superior in solidity to any cement. Even at this day Spain still
holds watch-towers that were erected by Hannibal.*"—Pliny's *Natural
History*, Bk. XXV, chapter xlviii.

J. St. Loe Strachey.

Newlands Corner,
Surrey.
1919.

CHAPTER ONE

General Survey

THERE is today a world shortage of almost every manufactured or cultivated product, as after the last war; there is also a labour famine and a lack of transport. In this country, closely connected with these deficiencies and looming ominously over them all, is our house famine. To relieve the last in face of the others, and without further aggravating them, is one of the most grave and pressing of the many problems that confront us.

Briefly the problem is this—to provide a maximum of new housing with a minimum expenditure of labour, transport, and manufactured materials.

For rural housing one of the solutions might be to use natural materials already existing on the site, materials which, without any elaborate or costly conversion, could, even by local labour, be worked straight into the fabric of the building. Also by such means it might be possible to build houses of the kind which would otherwise have a low priority in the post-war building programme.

Earth walling of one kind or another will usually fulfil the above conditions; it remains to choose what method shall be adopted in the particular circumstances, and to understand the limitations which attach to that method.

In principle all earth walling, except Chalk Rock which is properly a masonry construction, is based upon the fact that when certain suitable materials with the right moisture content are tightly compressed, they cohere to form a fairly hard, strong, and solid body. The means by which the necessary cohesion is obtained may be by external compaction of a relatively dry earth mix in shuttering (Pisé), or by the natural drying out of water from a wetter mix (Cob). In either case, once the process is complete, the cohesion lasts only so long as the materials are kept dry; if they should become thoroughly saturated, they will revert to mud and lose their cohesion and stability. At the outset, therefore, their limitations must be accepted: adequate protection against attrition by driving rain is of first importance—wide over hanging eaves, drips, damp-proof courses, and an efficient protective coating are a *sine qua non* of lasting earth walling.

Of recent years experiments have been made with cement or other agent as a stabilizer in order to obtain a stronger and more weather-resistant wall. This is given the general term Stabilized Earth.

CLASSIFICATION

A brief classification will help to give a clear picture of the traditional forms of earth walling and of their contemporary variations.

The techniques can be classified in the following way: Pisé de terre and rammed stabilized earths, both of which entail the use of shuttering; Cob and Chalk Mud, which are the traditional methods of building without shuttering; Earth Blocks, which include wet-moulded clay lumps made on a palette or tray, called Adobe or "sun-dried bricks," and also those blocks compressed in a retaining mould like Pisé, or in specially designed portable presses; Stabilized Earth blocks, which are similar to Pisé blocks, but with a stabilizing agent, and which are called Bitudobe when Bitumen is used; and Chalk Rock, known sometimes as "Clunch," which is a masonry block.

PISÉ

Pisé is a very old and simple method of building. The method consists in ramming earth in climbing shuttering, which is raised lift upon lift until the wall has reached its full height. This rude technique has continued to be practised from time to time all over the world, in Europe, Africa, and the Americas. Two- and three-storey dwellings can be found, particularly along the Rhône in France, where, with well-protected walls, many have stood for two and three hundred years.

Pisé is liable to reappear in history when bricklayers are scarce. This is understandable. When bricklayers are once more trained in numbers, the Pisé demand may again disappear, but it always has its uses, and, moreover, it is a *permanent* material.

STABILIZED EARTH

Stabilized Earth is the modern variation of Pisé. By the addition of cement or bituminous emulsion or other suitable agent the earth wall acquires greater strength and greater weather-resistant properties. The technique of stabilizing earths has been used fairly extensively in America for road-making and for building, and in view of its recent development it has been the subject of a more close scrutiny than have some of the traditional methods of earth building. There are therefore some scientific data which permit of more exact specification and enable the designer to make comparison with the standards of other methods of building.

COB

"Cob" building needs less introduction, as it is still well understood and was, until recently, a living craft in several parts of Great Britain, notably in Devonshire and South Wales, where its merits and advantages have been recognized apparently from earliest times.

CHALK MUD

"Chalk Mud" is a similar technique and has been practised until recently in parts of Wiltshire, especially in and around Winterslow, a village which has given its name to that particular method of walling. Both Cob and Chalk Mud walls differ from those of Pisé in the moisture and plasticity of the material, and in the absence of any shuttering. The wall is built up by the simple process of pitching on a soft but cohesive mixture of clay and straw, or chalk and straw, in layers, until the wall has reached its full height. The wall surfaces are pared down with a flat-backed spade to form a fair face as the work proceeds.

EARTH BLOCKS

The making of separate blocks, which is the alternative to monolithic construction, avoids the climbing shutter with its problems, but has the disadvantage of joints, and involves laying. Adobe, locally known as "Clay Lump," is found in this country in East Anglia, while similar Adobe buildings are common in the south-western States of America; the more recently developed Stabilized Earth blocks have also been used in America. Chalk Mud and Chalk Rock are found particularly in Wiltshire and adjoining counties. Chalk Rock has been used with effect in buildings of a monumental stature, such as Marsh Court in Hamp-

shire by Sir Edwin Lutyens, and in the Deanery Garden, Sonning
(Pls. 9 and 10), but, as has already been mentioned, Chalk Rock is
really a masonry construction and has therefore not been described in detail in this book.

Value of Earth Building

Those who are familiar with methods of earth walling are fully
alive to their virtues. This book, however, is addressed to those
who have in the past built only with stone, brick, concrete,
timber, or plaster, etc., and who are only now considering a rever-
sion to the more primitive constructions here described, through
the shortage or absolute lack of their former materials.

It is not so much a question as to whether you prefer a Cob or
Pisé house to one of brick or stone or concrete—though there are
some who profess a lively preference for the former—but as to
whether you will boldly revert to these old and well-tried methods
of building or, in the absence of the ordinary materials, feebly sit
down and build nothing at all.

For that is likely to be the alternative for a great many private
persons. National and Public-Utility Housing Schemes and
public and industrial works of all sorts will naturally and properly
claim priority in the matter of normal building materials and
skilled labour—and the private individual may not be able to
secure either for some time to come. In the case of normal materials he may have to pay the high price which is the logical outcome of a general shortage and of an unprecedented demand.

Timber, tiles, slates, plaster, and ironmongery he must still
purchase and transport as best he may—but the walls of his house
could be raised from the soil of the site itself by the employment of
some simple gear and a small amount of relatively unskilled local
labour.

If on suitable sites in rural areas earth-building techniques were
used for labourers' cottages and farm buildings, village halls,
garden houses, apple rooms, garages, and sheds, a contribution
could be made to the fulfilment of our present urgent building
needs. Skilled bricklayers would then be released for areas where
they were more urgently needed. Moreover, if such a policy
were resorted to, it would tend to check the importation of alien
materials to rural districts where their use is æsthetically
objectionable.

Pl. 9.—Marsh Court, Hampshire. A fine example of hewn-chalk building by Sir Edwin Lutyens.

Pl. 10.—Brick-and-chalk
vaulting at the Deanery
Garden, Sonning.

[17

GENERAL SURVEY

STRUCTURAL STANDARDS

It is sometimes asked whether houses of rammed earth can be built to conform to contemporary standards. So far as local authorities are concerned, walls of rammed earth have in a number of cases during the past few years been deemed to conform to the by-laws. Moreover, it is significant that the present values, for rating purposes, of some experimental cottages built at Amesbury in 1920 are not appreciably different from those of similar cottages in brick construction.

STRENGTH

So far as strength is concerned, all forms of earth construction have proved themselves adequate to support the loads of the normal two-storey house—about 1½ tons per foot run—provided walls of a generous thickness, usually about 18 in. to 2 ft., are used. These thicknesses could undoubtedly be reduced with some types of material if reliable figures were available as to compressive strengths, and if it were possible to build the walls under sufficiently controlled conditions. For instance, it is claimed that compressive strengths of 1000 lb./sq. in. and more can be achieved with cement stabilized earths, and while these figures may be somewhat extravagant, depending upon careful control in the soil mix and very thorough compaction, it should, nevertheless, be possible to exceed 400 lb./sq. in., at which strength it would be reasonable to use the material in cavity construction with leaves of 4 in. thickness, provided the usual number of wall ties were included. This would, of course, apply only to normal two-storey houses with joisted floors and with evenly distributed loading. Piers and narrow widths between openings carrying concentrated loads would naturally require to be stronger.

There are, therefore, two courses open. Questions of strength may either be virtually ignored by building walls of somewhat extravagant thicknesses, or they may be investigated scientifically by obtaining compressive strengths on samples, from which data it would be possible to arrive at more economical wall thicknesses.

RAIN PENETRATION

Experience has shown that earth walls can be made resistant to rain penetration, though the addition of suitable coverings is necessary to prevent surface erosion and to prevent penetration through any cracks in monolithic walls, and through the normal joints in block construction. As in brickwork, a cavity wall gives a greater protection against rain penetration than does a solid wall.

HEAT INSULATION

The heat-insulating properties of earth walls may vary considerably according to the type of earth used, the degree of ramming, and the amount of stabilizing agent which is used.

It is evident from the published figures that, with the exception of chalk, earth, as a walling material, has, contrary to popular belief, no great advantage in respect of heat insulation; earth walls probably owe their reputation as heat insulators to the fact that they were of necessity built thick, when they may have given better insulation than the traditional solid brick wall. Also, their greater mass would have a high heat capacity—that is, they would warm up and cool down slowly, which would help to maintain an even inside temperature with fluctuating weather conditions. There is a danger that in using thinner, stronger, walls this advantage of traditional earth walling may be lost. On the other hand, when intermittent heating is used in winter the characteristic slow rate of warming up of the mass walling may be a disadvantage and give rise to condensation. Under these conditions an internal lining such as a fibre board or a straw board would provide a surface which would readily adjust itself to temperature changes inside the house, and would, besides, add considerably to the heat-insulation value of the walls.

As a matter of interest some figures showing the heat-insulating properties are given. Figures for thermal conductivities, "k," quoted in a number of sources, suggest limiting values of about 11 and 3·5 B.Th.U./per hour/sq. ft./per in./per deg. F. difference in temperature for gravelly soils. Chalk is a better insulator, and conductivity values as low as about 2·4 B.Th.U./etc., have been given, though these may be exceptional. A 12-in. wall in chalk would thus have an overall thermal transmittance value $U = 0·17$ B.Th.U./per hour/per sq. ft./per deg. F. difference in temperature air to air. This is better than that of an 11-in. cavity wall in brick, which is about 0·3. But since thermal conductivity is related to the density of the material, and since the density of earth walls approximates to that of a gravel concrete, at about 130 lb./cu. ft., their thermal conductivity would be expected to be no more than that of the concrete; that is, about 7 B.Th.U./etc. For comparison, the thermal conductivity of brickwork may be taken as about 8 B.Th.U./etc.

Sound Insulation

Sound insulation in solid walls is also related to their density; here the denser the wall the better the insulation. Earth walls used as partitions must be about 6 in. thick to be satisfactory. It is nowadays recommended that in party walls a reduction in sound against air-borne noise should be not less than 55 decibels. A solid poured concrete wall of a concrete weighing not less than 80 lb./cu. ft., 9 in. thick, attains this standard when faced both sides with two coats of plaster on lathing fixed to battens. An earth wall of the same thickness, similarly finished, would be expected to be at least as good, and the normal 14-in. wall proportionately better.

Fire Hazard

Earth walls are normally considered to be incombustible, and resistance to fire is likely to be sufficient for dwelling-houses.

Durability

There can be no doubt that earth walls are durable; much, however, must depend upon the proper design of the house, and where permanent coverings to the walls, such as renderings or slate or tile hanging, are not used, upon the upkeep of the protective coating. Since many types of earth wall can be nailed into satisfactorily, there seems no reason why such slate or tile hanging should not be used in exposed positions in place of the usual treatments, although no examples can be quoted.

Vermin Infestation

The only other structural standard which need normally be considered is resistance to vermin infestation. There is no reason why houses of earth should not be as satisfactory in this respect as houses built of other materials, provided that proper attention is paid to detail in design. The normal criteria, such as that there should be a minimum of cracks and hollow spaces where bugs can enter and lodge without being detected, apply as much to other forms of construction as to earth building.

COST

It is unfortunately not possible to give much indication of the differences in cost between one type of earth building and another, and between earth building and other forms of construction. Much will depend upon the circumstances and, amongst other things, upon the availability of suitable material on the site and of labour, upon weather conditions at the time of building, upon the type and amount of stabilizing agent, if any, and upon the number of houses to be built. Clearly there would be no point in incurring the added expense of transporting suitable material from too great a distance if appropriate earth were not available on the site, or if other good building materials, with the necessary skilled labour, were available locally.

The expense of shuttering will be proportionately greater if only a single house is built, and since any saving in cost over other methods of building must be accounted for in the cost of the walls only, this item assumes some importance. Consideration might be given to hiring shuttering or, in some cases, to making use of the materials from the shuttering in the construction of the floors or roof.

However, there is evidence that under favourable conditions a saving over normal brick construction can be effected. For instance, the saving over local brick for three cottages built in Hampshire in 1913 was £54. In 1920 the cost of clay block houses built in Norfolk was £60 to £80 less than the estimated cost in brick for each house. Comparative costs of a number of different types of walling used in experimental cottages built at Amesbury in 1920 are as follows.[1]

[1] Report of the Department of Scientific and Industrial Research, *Experimental Cottages*, by W. R. Jaggard. H.M. Stationery Office, London, 1921.

GENERAL SURVEY

COMPARATIVE COSTS OF VARIOUS TYPES OF WALLING PER UNIT CUBE

Materials	Costs taken as
Brickwork in mortar with 2-in. cavity and galvanized-iron wall ties—11-in. work . .	10·0
Chalk and cement, 20 to 1, rammed between shuttering reinforced with wire netting every 3 ft. in height and rendered externally with one coat of lime slurry—15-in. work. .	8·1
Solid concrete, 8 to 1, rammed between shuttering—9-in. work	8·095
Chalk and cement blocks, 12 to 1, with 2-in. cavity, and galvanized-iron wall ties, and rendered externally with one coat of slurry—10-in. work	6·19
Wet chalk—"Winterslow" method—15-in. work	6·19
Rammed chalk Pisé rendered externally—15-in. work	5·95
Chalk and straw rammed between shuttering—15-in. work	5·0

A report[1] giving details of cost of walling of a number of other Amesbury cottages built in 1920 states that Pisé walls could be built at that time at a cost of 15s. per yd. cube as against 25s. per yd. super for 11-in. cavity brickwork. The figures were calculated for Pisé 18 in. thick on the lower floor and 14 in. on the first floor, with labour costs of 1s. 3d. per hour producing 1 ft. cu. of finished wall per hour, as against brickwork at £56 13s. 4d. per rod. The cost of scaffolding was not taken into account, but was thought to be less for Pisé than for brick.

When bricks and transport are expensive, the saving effected by earth walling may be significant, but in estimating differences in cost it should not be forgotten that although the cost of materials for earth walling is negligible, the total amount of labour required is proportionately greater than for other forms of construction.

[1] Report of the Experimental Cottages of the Ministry of Agriculture and Fisheries at Amesbury. *The Builder*, December 17th, 1920.

CHAPTER TWO

Pisé de Terre

Pisé DE TERRE" is merely the French for Rammed Earth, defined by Larousse as "Maçonnerie de terre argileuse comprimée sur place." In principle the method is a simple one involving nothing more than a certain amount of mechanical energy; for when particles of soil are forced together into close contact by ramming, they are compelled to adhere together by molecular and capillary forces and to form a hard and solid mass which is excellent for building purposes.

The odd thing is that the very obvious merits of Pisé should have secured it such small attention. It is no new-fangled war-time invention brought forth by our present necessity, but a very ancient system well proved by centuries of trial.

Pliny gives an excellent account in his *Natural History* of the massive watch-towers built in Spain by Hannibal during the second punic war. Remains of these structures are said to be standing to this day. Monsieur Gorffon, who published a treatise on Pisé in 1772, states that it was first introduced into France by the Romans. It has since spread throughout the remainder of Europe, to Africa, the Far East, and to the Americas, where dwellings built by the Indians are found along the Atlantic coast.

The following extracts from an old book based on a French original will serve well as an introduction to the study of Pisé building:

CAPABILITIES

"An account of a method of building strong and durable houses, with no other materials than earth; which has been practised for ages in the province of Lyons, though little known in the rest of France, or in any other part of Europe. It appeared to be attended with so many advantages, that many gentlemen in this country who employed their leisure in the study of rural economy were induced to make a trial of its efficiency; and the result of their experiments has been of such a nature as to make them

PISÉ DE TERRE

desire, by all possible means, to extend the knowledge and practice
of so beneficial an art.

"The possibility of raising the walls of houses two or even three
storeys high, with earth only, which will sustain floors loaded with
the heaviest weights, and of building the largest manufactories in
this manner, may astonish every one who has not been an eye-
witness of such things."

Of Pisé and Its Origin

"Pisé is a very simple manual operation; it is merely by com-
pressing earth in moulds or cases, that we may arrive at building
houses of any size or height."

Locale

"This art, though at present confined to the single province of
the Lyonese in France, was known and practised at a very early
period of antiquity. The Abbé Rozier, in his *Journal de Physique*,
says that he has discovered some traces of it (Pisé) in Catalonia;
so that Spain, like France, has a single province in which this ancient manner of building has been preserved. The art, however,
well deserves to be introduced into more general use. The cheapness of the materials which it requires, and the great saving of
time and labour which it admits of, must recommend it in all
places and on all occasions, but the French author says that it will
be found useful in hilly countries, where carriage is difficult, and
sometimes impracticable; and for farm buildings which, as they
must be made of considerable extent, are usually very expensive,
without yielding any return."

An Old Encyclopædia

There is an exhaustive article on Pisé in Vol. XXVII of *The
Cyclopædia or Universal Dictionary of Arts, Sciences, and Literature*, by
Abraham Rees, D.D., F.R.S., F.L.S., who draws chiefly on
French authorities. His directions are most detailed and precise,
and give an excellent picture of the practice then considered to
give the best results, when the book was published in 1819. Of
Pisé houses he says:

"Houses so built are strong, healthy, and very cheap, they will
last a great length of time, for the French author says he had pulled

23

down some of them which, from the title deed in the possession of the proprietors, appeared to be 165 years old, though they had been ill kept in repair. The rich traders of Lyons have no other way of building their country houses. An outside covering of painting in fresco, which is attended with very little expense, conceals from the eye of the spectator the nature of the building, and is a handsome ornament to the house.

"Strangers who have sailed upon the Rhône probably never suspected that those beautiful houses, which they saw rising on the hills around them, were built of nothing but earth, nay, many persons have dwelt for a considerable time in such houses without ever being aware of their singular construction. Farmers in that country generally have them simply white-washed, but others, who have a greater taste for ornament, add pilasters, window-cases, panels, and decorations of various kinds.

"There is every reason for introducing this method of building into all parts of the kingdom; whether we consider the honour of the nation as concerned in the neatness of its villages, the great saving of wood which it will occasion, and the consequent security from fire, or the health of the inhabitants, to which it will greatly contribute, as such houses are never liable to the extremes of heat or cold. It is attended with many other circumstances that are advantageous to the State as well as to individuals. It saves both time and labour in building, and the houses may be inhabited almost immediately after they are finished; for which purpose the holes made for the joists should not be closed up directly, as the air, if suffered to circulate through them, will dry the walls more speedily."

LOCAL TESTIMONY

At the end of the article, an instructive letter from a former rector of St. John's La Rochelle, is quoted:

"SIR,
    "My having been an inhabitant for some time of the town of Montbrison, capital of the Forêts, enables me to give you some information concerning the mode of building houses with earth, etc.
    "The church was the most remarkable in this style of building;

it is about 80 ft. long, 40 ft. broad, and 50 ft. high; the walls built of pisé, 18 in. thick, and crêpé, or rough-cast on the outside, with lime and sand. Soon after my arrival, the church, by some accident, was destroyed by fire, and remained unroofed for about a twelvemonth, exposed to rains and frost. As it was suspected that the walls had sustained much damage, either by fire or the inclemency of the season, and might fall down, it was determined to throw them down partially, and leave only the lower parts standing; but even this was not done without much difficulty, such was the firmness and hardness these walls had acquired, the church having stood above eighty years; and all the repairs required were only to give it on the outside, every twelve or fifteen years, a new coating of rough-cast.

"A house for a single family is generally finished in about a fortnight. The following is the method I have seen them practise:

"The earth is pounded as much as possible, in order to crumble any stones therein; clay is added thereto in a small quantity, about one-eighth part. It is all beaten and mixed up together by repeated blows with a mallet about 10 in. broad, and 10 or 15 in. long, and 2 in. thick. The earth being thus prepared, and slightly wetted, the foundation of the house is dug for; this is laid with stone, and when it is about 1 ft. high above the surface of the ground, planks are arranged on each side, which are filled with earth intended for the wall; this is called Pisé in the dialect of the country. It is strongly beaten; and this method is continued successively all round the building. The walls have more or less thickness according to the fancy of the owner; I have seen them 6 in. and 18 in. thick. If the several storeys are intended in such erections, they do not fail to place beams to support the floors before they build higher. Of such buildings I never saw any consisting of more than three floors at most; generally they have but two. When the building is thus finished, it is left for some months to dry; then such as wish to make the building more solid and durable, give it a rough-cast coating on the outside with lime and sand. This is what I have observed during a residence of three years in the town of Montbrison. I should be happy if this detail should afford the slightest information to the generous nation which has received us with so much goodness.

"I am, etc.,

"JACOUR."

COB, PISÉ, AND STABILIZED EARTH

THE TECHNIQUE OF BUILDING IN PISÉ

SINCE the methods of Pisé building today do not differ materially from the traditional practice, and, except in the more scientific knowledge which now exists on the behaviour and uses of soils, are but amendments to the old system, it may be as well to consider, side by side with those of present-day practice, the principles set forth in the following extracts from the article by Dr. Rees in the encyclopædia mentioned.

Dr. Rees introduces his subject thus:

"Pisé building, in Rural Economy, the name of a method of building with loamy or other earthy matters, which has long been practised with great success, and in a very cheap manner, in some departments of France, and which is now had recourse to with similar advantage in some parts of this country. It has been described, delineated, and recommended by Mr. Holland [1] in the first volume of *Communications to the Board of Agriculture*, and is to be managed somewhat in the manner directed below."

At great length, and with immense detail, the plant, the preliminaries, and the process are each severally described, but the pith of the matter is sufficiently given by the following extracts:

SHUTTERING AND RAMMERS

"The instrument with which the earth is rammed into the mould is a tool of the greatest consequence, on which the firmness and durability, in short the perfection, of the work depends. It is called a pisoir, or rammer; and though it may appear very easy to make it, more difficulty will be found in the execution than is at first apprehended. A better idea of its construction may be formed by examining the Plate (Pl. 19) in which it is delineated, than any words can convey. It should be made of hard wood, either ash, oak, beech, walnut, etc., or what is preferable, the roots of either of them.

"For the construction of the mould, take several planks, each 10 ft. long, of light wood, in order that the mould may be easy to handle; deal is the best as being least liable to warp. To prevent which the boards should be straight, well seasoned, and with as

---

[1] The "Mr. Holland" referred to, is of course Henry Holland, the architect of Carlton House and Brooks's Club; it is interesting to note that this eminent architect should have taken an interest in Pisé.

26

few knots as possible. Let them be well ploughed and tongued and planed on both sides. Of these planks, fastened together with four strong ledges on each side, the mould must be made, 2 ft. 9 in. in height; and two handles should be fixed to each side.

"All the boards and ledges here mentioned must be, after they are planed, something more than 1 in. thick."

Opinion varies today as to the most suitable type of rammer, and as to the necessity for using one which is heart-shaped. Modern hand rammers are sometimes made of cast iron, with a shaft of 1-in. galvanized pipe forming the handle. For the Amesbury experiment (see Chapter VIII), three different types were tried—a flat iron rammer, a large wooden one of heart shape, and a smaller heart-shaped one of iron which was found generally to be the most satisfactory. The flat rammer tended to "case harden" the top surface of the layer being rammed, and workmen complained that the large wooden rammer tended to pick up the material. The weight of the rammers was between seven and twelve pounds.

The use of pneumatic rammers (see Pl. 11) may well prove to be of economic advantage if they can be rented at a reasonable cost, and if there is a fairly extensive amount of walling to be done. A pneumatic rammer should triple the rate of walling achieved by hand ramming.

Mr. Francis Macdonald, who has carried out research in America on building in Pisé and in Stabilized Earth, discusses different types of rammer in his booklet *Terracrete*.[1]

"Hand rammers are simple, consisting merely of a fairly heavy head to which a handle is firmly attached. Individual's requirements will govern somewhat as to size and weight; but most preferences will be satisfied within the following limits: Face area, 9 to 16 sq. in. for a total weight between 15 and 20 lb.; and from 5 to 6 ft. for a handle of 1-in. galvanized pipe. (A solid iron head will weigh approximately ¼ lb. per cu. in.; 1-in. pipe, 1·7 lb. per ft.) A rammer too light for its face area requires extra and eventually tiring effort to bring it down with sufficient force enough for proper compacting, while an overweight rammer is wearying to lift repeatedly.

"Concerning heads, there seems little practical or theoretical

[1] Francis Macdonald, *Terracrete—Building with Rammed Earth Cement*. Chestertown, Maryland, U.S.A., 1939.

reason for the wedge and heart shapes that are frequently recommended. They neither compact material directly nor permit proper working along the sides and in the corners of forms. The rammer head may be of iron, steel, or wood, steel shod. (To adjust weight and balance of a steel-shod head, the pipe handle may be loaded.) It should be square or rectangular in cross section, the metal face smooth and flat. Edges and corners should be filed slightly to reduce scarring of forms, but only slightly; too great a bevel or rounding tends to pack earth against the form sides and increase sticking. The handle should be carefully perpendicular to the bottom face and firmly attached, preferably by welding or brazing."

The width or diameter of the head of an air-hammer "should be equal to or greater than the diameter of the barrel above, to permit working along form sides and corners, because striking at an angle may cut forms badly. The weight should be approximately that of the butt supplied with the hammer. Increase in weight will slow the striking, and vice versa. For this reason a larger butt, with greater face area, can be made of aluminium with a consequent increase in walling rate."

The plant commonly used for Pisé building until the time at which the first edition of this book was published was but a slight improvement on the old traditional models, as can be seen by comparison with old engravings and descriptions. Pisé building lay off the great main stream of constructional activity, and the enterprise and ingenuity lavished on the perfecting of other building materials and methods passed Pisé by, leaving it undisturbed in its quiet backwater, a primitive system still with its primitive tackle.

In the accepted shuttering there were a number of very obvious and unnecessary shortcomings which seemed to clamour for attention; defects, too, that were in no way inherent, but merely traditional infelicities reproduced in succeeding models which remained remarkably true to their primitive ancestral archtype—the Pisé plant described by Pliny.

The chief desiderata in designing satisfactory Pisé shuttering appear to be these:

All constituent parts should be reasonably light and easy to handle. The shutters should be rigid and not liable to warp, and should not be expensive to construct. Since the outward thrust developed during ramming is very great, they should be constructed with considerably more rigidity than those used for concrete construction, and any tendency to spring must be checked. When clamped in position the shutters should be firmly and positively supported, without deviation from the vertical.

The shuttering must be rapidly and easily disengaged and removed from the wall.

The fairway between the shutters must be as little obstructed by the cross-braces as may be, leaving good room for the men on the wall to tread and ram.

The through-pins by which the shuttering rests upon the base wall or on a completed course of Pisé must be easily withdrawn without injury to the wall.

There must be either a special corner form, or some device by which the ordinary shutters may be rigidly clamped together to form the corner.

There must normally be some means of blocking off the shuttering at any desired point, for the forming of door or window openings at any level, although with the use of a plan with standard sized openings and standard spaces between them, this difficulty may be largely overcome. Reference is made to this point under "Walls" in Chapter VII.

If the partition walls are to be built in Pisé, some provision must be made for adjusting the shuttering to a smaller thickness of wall, unless another set of shutters is to be used.

The whole apparatus must be as simple and as fool-proof as possible, and built to stand rough usage and exposure to the weather.

In 1920 the author attempted to construct a system of shuttering embodying these essentials, and the working drawing (Fig. 1) will give the reader a tolerable idea of what was called the "Mark V" model. The principle of the building process remained unaffected, as indeed it has in all subsequent types of shuttering so far.

The Mark V shuttering was tested in one of the London parks, and the trials were sufficiently satisfactory to encourage a belief that it would prove to be a very considerable improvement on the old traditional types. It was then despatched to Surrey to undergo the searching and very practical test of being used to construct a small-holder's house and homestead. A description of this house, which was built at Newlands Corner, is given on pages 53–58.

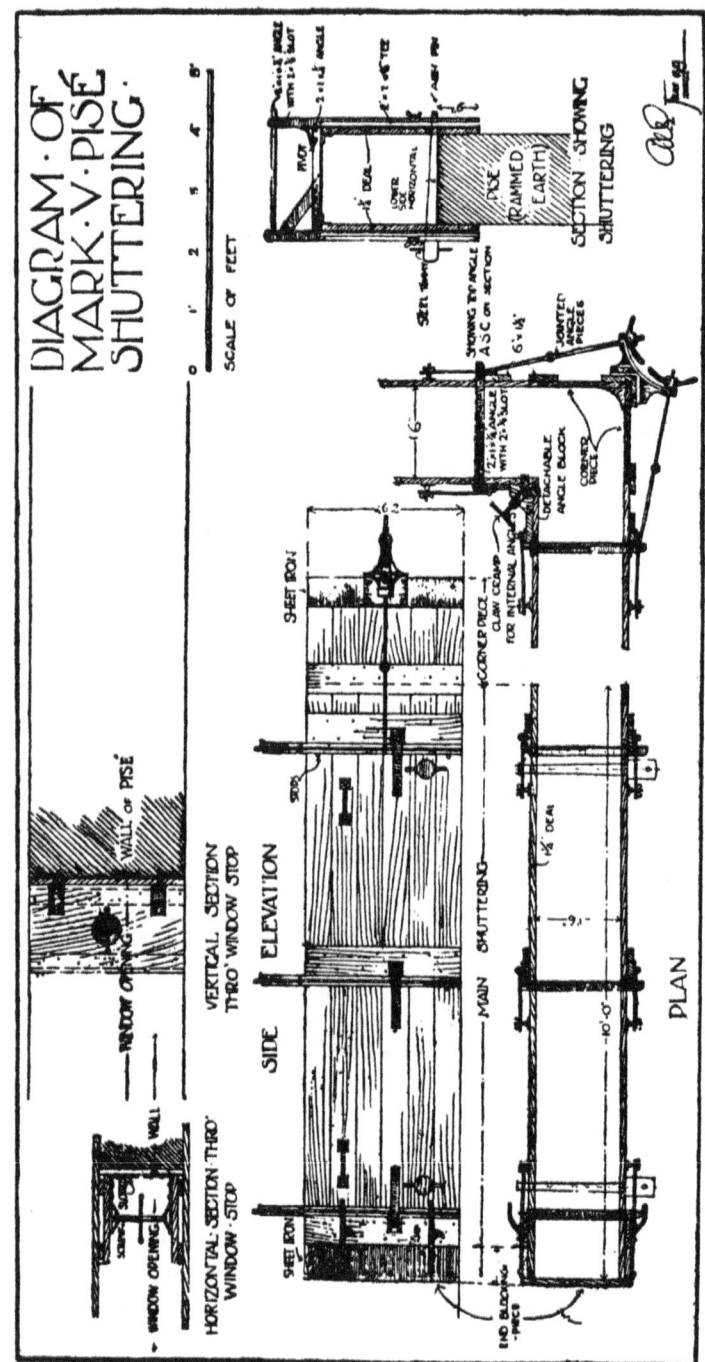

DIAGRAM · OF
MARK · V · PISÉ
SHUTTERING.

SCALE OF FEET

SECTION SHOWING SHUTTERING.

VERTICAL SECTION THRO' WINDOW STOP

WALL OF PISÉ

WINDOW OPENING

HORIZONTAL SECTION THRO' WINDOW STOP

WINDOW OPENING

WALL

SIDE ELEVATION

SHEET IRON

CORNER PIECE
CLAW CLAMP
FOR INTERNAL ANGLE

DETACHABLE ANGLE BLOCK

CORNER PIECE

MAIN SHUTTERING

END BLOCKING PIECE

SHEET IRON

PLAN

Fig. I.

30

Since that time various experiments have been made to produce a type of shuttering weighing considerably less than the traditional kind, including, for instance, those made with ¾-in. plywood by Mr. Macdonald. The results obtained from these were unsatisfactory, since the bracing required to keep the plywood rigid took too long in assembly, and added too much to the weight of the shutter to make its use worth while. Plywood was also found to be insufficiently proof against continual resetting and scarring from the rammers.

There are obviously ways in which sheet metal could be used, either flat or corrugated, or by combining the two, which would give a light and rigid shutter. Here seems to be a field for an ingenious inventor, a field which is still "To Let." There are, it is true, patent steel shutters, such as the patent climbing shuttering produced by Joseph Bradbury & Sons Ltd., of Braintree, and illustrated in Pls. 12 & 13, which have been used for the construction of houses of semi-dry concrete. Shutters of this type, which have a free fairway without bolts or struts, and which could be hired, might prove ideal for Pisé work, though to the authors' knowledge they have never been used for this purpose. The possibility of building cavity earth walls with them, as in concrete, is also worth considering.

Other experiments appear to have been confined to the development of two distinct types of shutter. One is a shallow type which is braced by clamps or yokes fitting over the top, and which gives a course of about twelve inches high; the other is a deeper type which gives a course of about thirty-six inches in height, and which is braced either by through bolts which must be withdrawn when the shutter is dismantled, or by pieces of wire which are cut off and left in the wall (Fig. 2). The former type is reported to be quicker to handle on account of its lightness, if it is well designed, and more than three courses may be made with it in the time in which it takes to make one course with the deeper type.

A typical set of wooden climbing shutters is made 10 to 12 ft. long. The timber used is 1½ in. in thickness, and requires to be stayed at 30-in. intervals. The drawings show the method of construction of these types of shutter better than can be described in words. Even where shorter lengths of shuttering are constructed, for use on short runs of internal wall, the timber will

SHUTTERS OF DEEP AND YOKE TYPES

Ⓐ DEEP TYPE WITH THROUGH BOLTS

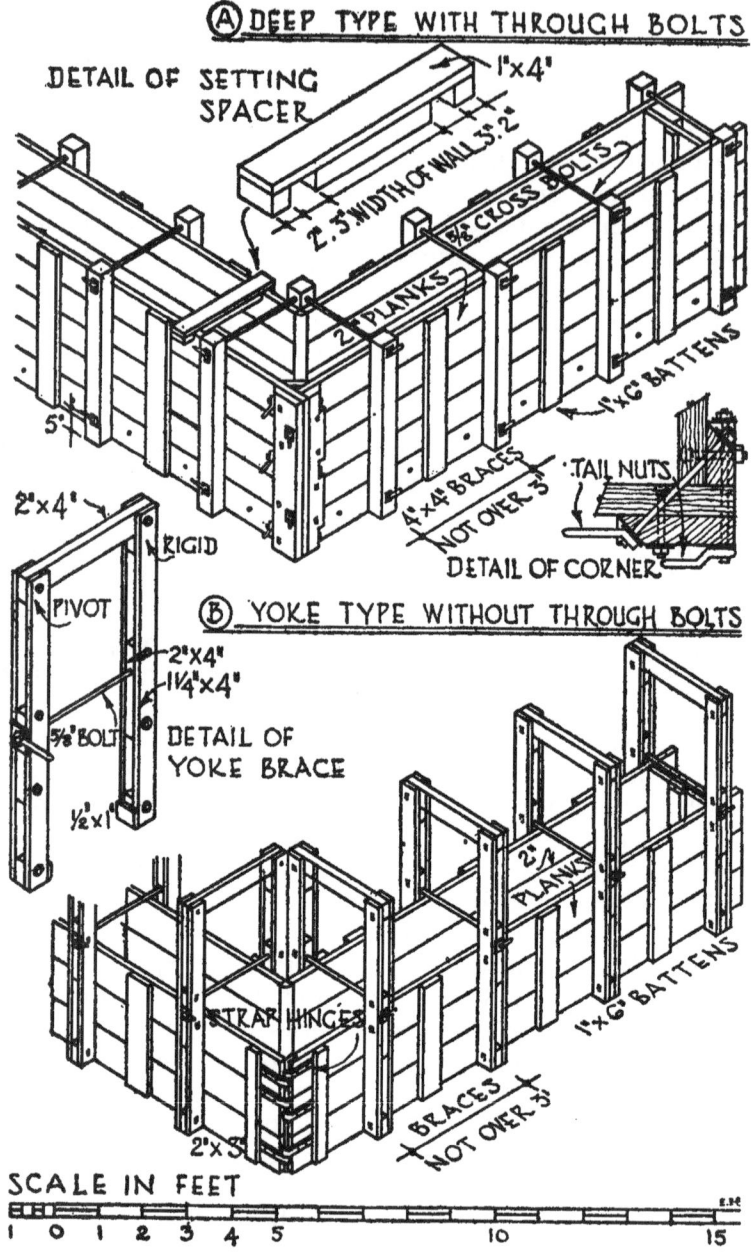

DETAIL OF SETTING SPACER

1"x4"

2.5 WIDTH OF WALL 3".2"

⅜ CROSS BOLTS

2" PLANKS

1"x6" BATTENS

5"

4"x4" BRACES

NOT OVER 3'

TAIL NUTS

DETAIL OF CORNER

2"x4"

RIGID

PIVOT

2"x4"

1¼"x4"

⅝" BOLT

DETAIL OF YOKE BRACE

½"x1"

Ⓑ YOKE TYPE WITHOUT THROUGH BOLTS

2" PLANKS

1"x6" BATTENS

STRAP HINGES

BRACES NOT OVER 3'

2"x3"

SCALE IN FEET

1 0 1 2 3 4 5   10   15

Fig. 2.

32

Pl. 11.—Ramming the first course of Pisé walls with air hammers.

Pl. 12.—Joseph Bradbury & Sons, Ltd., G.E. Climbing Steel Shuttering, Lifters and Clamps.

Pl. 13.—G.E. Patent Shuttering, showing internal clamps, cavity cone, and use of shutters for carrying internal scaffold.

still need to be 1½ in. in thickness since the distance between the
uprights remains the same.

Fig. 3 shows a new type of shuttering designed by Mr. Ernst
May. The shutters, which have been used for building in
Nairobi, are held in position by adjustable cramps which can be
used for the making of walls from 9 in. to 24 in., as can be seen from

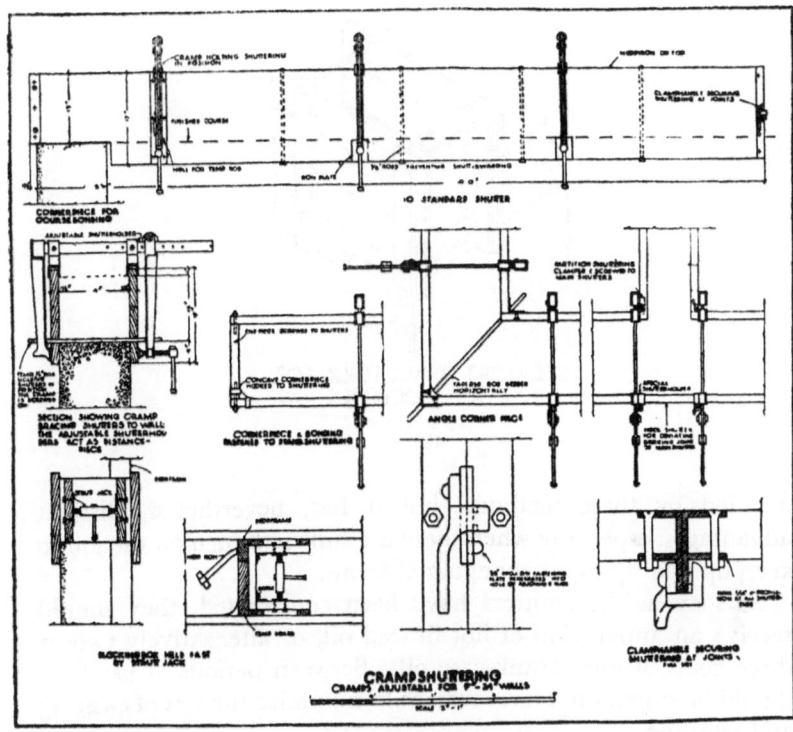

Fig. 3.

the drawing. This sort of shuttering has the advantage that it
can easily be put in position and moved by unskilled labour.

There are three ways in which corners may be made. One
method is to fix an end board on by means of cleats to one of the
straight wall forms, the bottom boards of which must be cut away
a depth of 6 in. and a length equivalent to the thickness of the
walls, so that the shutter may grip the wall being built, and at the
same time extend over the intersecting wall (Fig. 4). The second

method is to construct a clamping device (Fig. 27), whereby two long wall forms may be clamped together to form the corner. The disadvantage of this is that the corner must be detached whenever the forms are required to make a straight run of wall. The third method, that of using a special corner shutter, can be

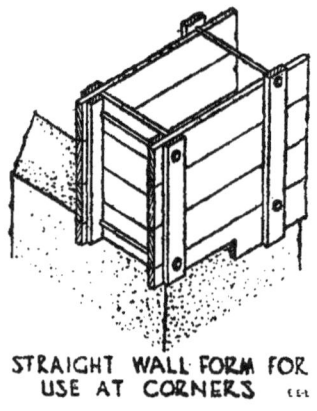

STRAIGHT WALL FORM FOR
USE AT CORNERS

Fig. 4.

avoided by these methods, but it has, nevertheless, distinct advantages, especially when there are sufficient men on the job to keep up a fairly rapid changing of forms.

As soon as the shutters have been constructed, they should receive an application of hot linseed oil, or alternatively two or three coats of used crank-case oil. Between periods of use they should be stored on a level surface to minimize the risk of twisting and warping.

The bolts holding the bottom of the shutter together may be withdrawn more easily if a layer of sand be placed round them before the earth is rammed on top. The sand may be poked out afterwards and the holes filled in with earth mortar. These bolts, which may be $\frac{3}{4}$ in. wide and threaded at each end, should be fitted with 3- to 4-in. lever arm tail nuts, to avoid having to use a wrench for dismantling them. If washers be screwed to the outside of the forms, the nuts do not work into the wood and wear through it.

A good method of forming a bond between one wall section and

34

the next is to fix a bevelled fillet to the end stop of the form, as shown in Fig. 5.

Fillets should be nailed to the shuttering at the corners, so that the sharp arrises of the building are rounded off to prevent their being damaged.

Under some conditions it is possible that the floor and roof timbers destined for use in the house under construction may be

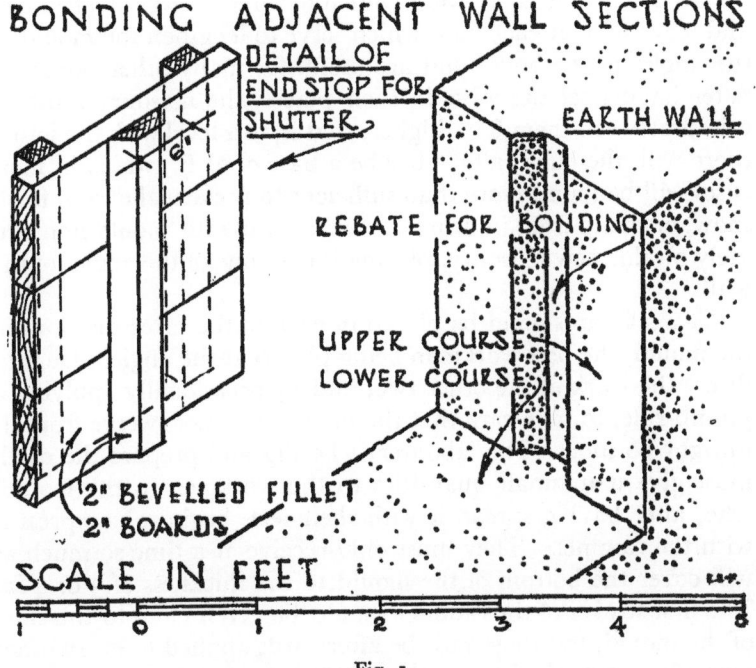

Fig. 5.

used economically and satisfactorily as temporary shuttering for the earth walls. A Pisé "Test-house" was built in this way by Messrs. Alban Richards at their Ashtead works, and proved to be highly successful.

METHOD OF WORKING

Of the method of working, the encyclopædia says: "Pisé contains all the best principles of masonry, together with some rules peculiar to itself, which are now to be explained.

"To begin with the foundation; this may be made of any kind of masonry that is durable, and should be raised to the height of 2 ft. above the ground; which is necessary to secure the walls from the moisture of the earth, and the splashing of the rain which will drop from the eaves of the roof. When these foundation walls are made level, and 18 in. thick, mark upon them the distance at which the joists are to be set, for receiving the moulds; those distances should be 3 ft. each from centre to centre. Each side of the mould being 10 ft. long, will divide into three lengths of 3 ft. each, and leave 6 in. at each end, which serve to lengthen the mould at the angles of the house and are useful for many other purposes. After having set the joists in their places, the masonry must be raised between them 6 in. higher, that is, to a level with the joists; there will, therefore, altogether be a base of 2½ ft., which in most cases will be found more than sufficient to prevent the rain, frost, snow, or damp from injuring the walls. Raise the mould immediately on this new masonry, placing it over one of the angles of the wall.

"A workman should be placed in each of the three divisions of the mould, the best workman being placed at the angle. He is to direct the work of the other two, and by occasionally applying a plumb-rule, to take care that the mould does not swerve from its upright position. The labourers who dig and prepare the earth must give it in small quantities to the workmen in the mould, who, after having spread it with their feet, begin to compress it with the rammer. They must only receive at a time so much as will cover the bottom of the mould to the thickness of 3 or 4 in. The first strokes of the rammer should be given close to the sides of the mould, but they must be afterwards applied to every other part of the surface; the men should then cross their strokes, so that the earth may be compressed in every direction. Those who stand next to one another in the mould should regulate their strokes so as to bear at the same time under the cord, because that part cannot be got at without difficulty, and must be struck obliquely; with this precaution, the whole will be equally compressed. The man at the angle of the wall should beat carefully against the head of the mould.

"Care must be taken that no fresh earth is received into the mould till the first layer is well beaten, which may be ascertained by striking it with a rammer; the stroke should leave hardly any print on the place. They must proceed in this manner to ram in layer after layer, till the mould is full. When this is done, the machine may be taken to pieces, and the earth which is contained will remain firm and upright, about 9 ft. in length and 2½ ft. in height. The mould may then be replaced for another length, including 1 in. of that which has first been completed.

"The first course being thus completed, we proceed to the second; and here it must be observed that in each successive course we must proceed in a direction contrary to that of the preceding. It may easily be conceived, that with this precaution the joints of the several lengths will be inclined in opposite directions, which will contribute very much to the firmness of the work. There is no reason to fear overcharging the first course with the second, though only just laid; for three courses may be laid without danger in one day.

"This description of the first two courses is equally applicable to all the others, and will enable any person to build a house, with no other materials than earth, of whatever height and extent he pleases.

"With respect to the gables, they may be made without any difficulty, by merely making their inclination in the mould and working the earth accordingly.

"In one single day three courses of about 3 ft. each may be laid one over the other; so that a wall of earth of about 8 or 9 ft., or one storey high, may be safely raised in one day. Experience has proved that as soon as the builders have raised their walls to a proper height for flooring, the heaviest beams and rafters may without danger be placed on the walls thus newly made; and that the thickest timber of a roof may be laid on the gables of Pisé the very instant they are completed.

"*Soil Preparation.*—All the operations of this art are very simple and easy; there is nothing to be done but to dig up the earth with a pickaxe, break the clods with a shovel, so as to divide it well, and then lay it in a heap, which is very necessary, because as the labourers throw it on that heap, the lumps of earth and large stones roll to the bottom, where another man may break them or draw them away with a rake. I must observe that there should be an interval of about an inch and a quarter between the teeth of the rake, that the stones and pebbles of the size of a walnut or something more may escape, and that it may draw off only the largest. If the earth that has been dug has not the proper quality, which is seldom the case, and it is necessary to fetch some better from a distance, then the mixture must be made in this manner: one man must throw one shovelful of the best sort, while the others throw five or six of the inferior sort on the heap, and so more or less according to the proportions which have been previously ascertained.

"No more earth should be prepared than the men can work in one day, or a little more, that they may not be in want; but if rain is expected, you must have at hand either planks, mats, or old cloths to lay over the heap of earth, so that the rain may not wet it; and then as soon as the rain is over, the men may resume their work, which, without this precaution, must be delayed; for it must be remembered that the earth cannot be used when it is either too dry or too wet, and, therefore, if the rain should wet it after it has been prepared, the men will be obliged to wait till it has recovered its proper consistency—a delay which would be equally disadvantageous to them and their employer. When the earth has been soaked by rain, instead of suffering compression, it becomes mud in the mould; even though it be but a little too moist, it cannot be worked; it swells under the blows of the rammer, and a stroke in one place makes it rise in another. When this is the case, it is better to stop the work, for the men find so much difficulty that it is not worth while to proceed. But there is not the same necessity of discontinuing the work when the earth is too dry, for it is easy to give it the necessary degree of moisture; in such a case it should be sprinkled with a watering-pot, and afterwards well mixed up together; it will then be fit for use.

"It has already been observed that no vegetable substances should be left in the earth; therefore in digging, as well as in laying the earth in a heap, great care must be taken to pick out every bit of root, great and small, all sprigs and herbs, all bits of hay and straw, chips of shavings of woods, and in general everything that can rot or suffer a change in the earth.

"*Speed of Building.*—Besides the advantages of strength and cheapness, this method of building possesses that of speed in the execution. That the reader may know the time that is required for building a house, or an enclosure, he need only be told that a mason used to the work can, with the help of his labourers, when the earth lies near, build in one day 6 ft. square of the Pisé."

Some further comments are necessary in the light of modern requirements and techniques on the three aspects of Pisé building just described; that is, on the method of working, the soil preparation, and the speed of building.

The foundations should be built of concrete, brick, or rubble, whichever material is the cheapest and the most readily available in the locality. They should be built to a depth suitable for a masonry or brick wall of equivalent thickness, and to a height of between 1 and 2 ft. above ground level. If it is possible, a water table, either in the form of concrete *in situ*, or of stone or concrete flags, should be placed round the building, draining outwards, to divert the water which drops from the eaves away from the walls; especially should this be done in the case of a thatched roof having no gutters. By being raised a foot or so from ground level, the Pisé walls are protected from splashing, as stated in the encyclopædia, and risk of damage to the base of the wall is thereby minimized.

Since the time at which the *Cyclopædia* was compiled, the inclusion of a damp-proof course in anything other than sheds and poultry houses, etc., has become universal, and indeed under the by-laws obligatory. Most of the standard types are suitable for Pisé work, but if slates or other materials which may be shattered are used, they should have at least two courses of brickwork laid above them for protection, to prevent their being fractured when the earth is rammed in place above them. If the foundations are of brick, two courses of blue bricks in cement mortar will serve well as a damp-proof course. Tarred paper, bituminous felt, and also asbestos shingles in mortar have been used, but with the latter, care must be taken, as with slates, to prevent fracture.

Earth walls are vulnerable to the borings of rats and, in some countries, to attack by ants. Some precaution should be taken against these pests in those localities where they are a nuisance. The most efficient way of dealing with rats is to ram broken glass in with the earth for the first two feet or so of the Pisé wall. In tropical countries, carbolinum, arsenite of soda, or other ant-preventive is sometimes mixed with the earth of the bottom course to prevent white ants from doing damage to the walls. Other methods of excluding ants are suggested in Chapter IX.

If it can be arranged, the soil which is to be used should be prepared and kept under cover so that enough of it, of the correct moisture content, is available to allow the men who are ramming to be well supplied at the beginning of the day. As has already been described, the large pebbles should be removed in preparing the earth. An alternative method to raking them away as described in the encyclopædia is to throw all the earth hard against a ¾-in. screen.

With a good earth the thickness of wall might generally be 14 in. for a one-storey building, and 18 in. for a structure of two storeys. Internal partitions, which should be well bonded to the outside walls, should be about 9 to 12 in. thick, the latter if they are load bearing. When there is sufficient timber, partitions are sometimes studded at about 30-in. intervals, with rammed earth panels placed between the studs. Planks should be clamped to the studs to form the shuttering, and a triangular fillet should be nailed to them to form a bond with the earth panels. This pier and panel method has also been used for external walls, but care must be taken to ensure that the vertical joints with the piers are so designed that they remain watertight even if shrinkage takes place in the panel.

Sharp arrises on external walls are easily damaged, and the

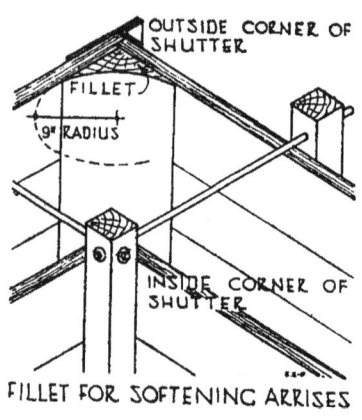

FILLET FOR SOFTENING ARRISES

Fig. 6.

corners of a building should be rounded off to a radius of about 9 in. or should be splayed. Fillets may be inserted in the corner moulds (Fig. 6), for this purpose.

The corners may be strengthened by using quoins of brick,

stone masonry, or concrete, but these are not always successful because of the shrinkage and cracking which results from the difference in volume change, which occurs during the drying out of the wall between the material of which the quoins are made and the earth walls. This causes cracks to form at the point of contact of the two materials and consequent moisture penetration at that point. A more successful practice is to block out spaces for the quoins during ramming and to build in the quoins after the earth wall has dried out and shrinkage has already taken place.

On beginning ramming, the mould should be levelled up, and a workman placed in each division, with the most skilful at the corners. It is important not to allow the mould to get out of place at the beginning, as it is most difficult to correct its position when it is half full. Heavy blows are unnecessary, and except for ramming the earth under the bolts, the men should be instructed not to work in unison, or too great a strain will be put upon the shuttering. A ringing sound from an iron pisoir will announce that the earth is well compacted, and the rammer will then make very little impress on it. When compacting the top layer in the mould, the surface should be indented with a heart-shaped pisoir, and should be sprinkled with water when the next layer is laid on top, in order to afford a good bond between the two lifts. The position of the mould in each course should be so arranged that no straight joints occur between one course and the next immediately above or below it. It is said to be easier to bond the vertical joints by means of a fillet in the stop end of the shuttering as described above, than by the method given by the *Cyclopædia* of sloping the earth down at 45 degrees at the end of the mould.

Wooden plugs and plates should be inserted in the mould as the work proceeds, so that the earth may be rammed round them. The heights of the lifts of the shuttering should be so arranged that the walls are first rammed to about 6 in. above the top of the final position of the joists and are then chiselled out to receive the joists (Fig. 7), when the shuttering is removed. In this way the joists may be inserted easily and sufficient earth will remain above them for the shuttering to rest on when the next course is rammed. The chiselling out should be done as soon as possible after ramming, as the earth quickly becomes very hard.

Lintols should be supported firmly by strutting after they have been put in place, so that they may withstand the pressure caused by subsequent ramming.

The speed of building quoted in the *Cyclopædia* appears to remain valid today, for it has been found that three men can easily

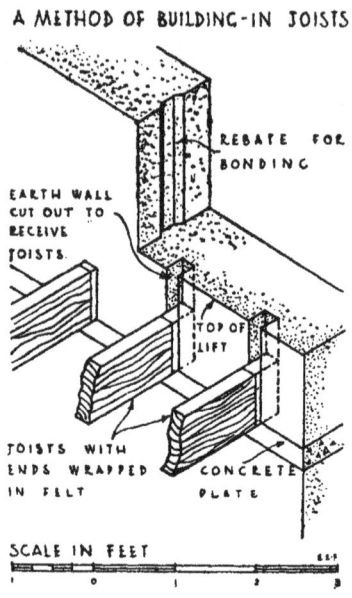

A METHOD OF BUILDING-IN JOISTS

Fig. 7.

place 2 cu. yds. of wall in a day by hand ramming; in fact, it is found that they may need to be supplied with earth sufficient for about 4 cu. yds.

REINFORCEMENT

This was used occasionally in traditional construction, but today, with a reasonably good soil, and with a design which recognizes the limitations which Pisé imposes, it is usually found to be unnecessary. Occasionally, however, when using a rather poor soil, or when the construction of narrow piers between openings is necessary, the earth walls can be reinforced. Wood was used traditionally, but in America scrap-iron has been used. In some laboratory tests made there, metal lathing, barbed wire, and metal rods were found materially to increase the bending moments of earth beams when subjected to point loads. Where reinforcement is necessary metal may have to be used in this country while there is a shortage of timber. However, there is very little evidence of the effectiveness of reinforcement of this kind in practice, and the only guide is the traditional method of timber reinforcement described in the *Cyclopædia* as follows:

"To make good walls, it is not sufficient that the earth be well beaten, we must also learn to unite them well together. Here the binders cost very little; they consist only of thin pieces of wood, a few cramps and nails, and these are sufficient to give the greatest stability to buildings of Pisé."

Having gone on to explain that the angles of the building are formed by the successive courses alternately crossing one another on the corners like the alternating "long and short" quoins in a stone building, our authority proceeds to describe how tough boards are laid between the courses of Pisé so as to cross at the corner and so, entirely encased in tightly compressed earth, form effective ties.

"This board must be rough, as the sawyers have left it, 5 or 6 ft. long, something less than 1 in. thick, and in breadth about 8, 9 or 10 in., so that there may remain on each side 4 or 5 in. of earth, if the wall is 18 in. thick; by this means the board will be entirely concealed in the body of the wall. When thus placed neither the air nor damp can reach it, and of course there is no danger of its rotting. This has often been proved by experience, as in taking down old houses of Pisé such boards have always been found perfectly sound, and many that had not even lost the colour of new wood. It is easy to conceive how much this board, from the pressure of the work raised above it, will help to bind together the two lengths of wall and to strengthen the angle.

"It is useful (particularly when the earth is not of very good quality) to put ends of planks into the Pisé after it has been rammed about half the height of the mould. These ends of planks should only be 10 or 11 in. long, to leave as before a few inches of earth on each side of the wall, if it is 18 in. thick; they should be laid cross-wise (as the plank before mentioned is laid lengthwise) over the whole course, at the distance of about 2 ft. from one another, and will serve to equalise the pressure of the upper parts of the works on the lower course of the Pisé.

"The boards above mentioned need only be placed at the

angles of the exterior wall, and in those parts where the courses of the partition walls join to those of the exterior wall, the same directions that have here been given for the second course must be observed at each succeeding course, up to the roof. By these means the reader will perceive that an innumerable quantity of holders or bondings will be formed, which sometimes draw to the right, sometimes to the left of the angles, and which powerfully unite the front walls with those of the partitions; the several parts deriving mutual support from one another, and the whole being rendered compact and solid.

"Hence these houses, made of earth alone, are able to resist the violence of the highest winds, storms and tempests. The height that is intended to be given to each storey known, boards of 3 or 4 ft. in length should be placed beforehand in the Pisé, in those places where the beams are to be fixed, and as soon as the mould no longer occupies that place, the beams may be laid on, though the Pisé be fresh made; little slips of wood, or boards, may be introduced under them, in order to fix them level. The beams thus fixed for each storey, the Pisé may be continued as high as the place on which you intend to erect the roof."

FIREPLACES AND FLUES

Fireplaces and flues are seldom constructed of Pisé, for amongst other reasons considerable skill is required to ram them successfully. It is found better to construct the stacks in brickwork or masonry, than to attempt to ram them in earth. Some of the precast chimney blocks and flue liners now on the market could be used with advantage in Pisé building.

RENDERING

Although rammed earth walls have been known to stand for over two hundred years in some climates without any permanent protective covering, the average Pisé wall cannot be expected to last long in wet climates without suffering attrition as a result of continual dampness and driving rain, unless it is given some coating of a protective kind externally which is permanently and properly maintained. This problem will be dealt with more fully in Chapter VI which is devoted to renderings.

PISÉ DE TERRE

DESIGN

In addition to those design points already mentioned in this chapter, there are others which are common to all types of earth building, and which have therefore been grouped together into one chapter (see Chapter VII).

SOIL MIXES

The *Cyclopædia* describes in a most picturesque manner the most suitable soils to be used for rammed earth building:

"1st. All earths in general are fit for that use, when they have not the lightness of poor lands nor the stiffness of clay.

"2ndly. All earths fit for vegetation.

"3rdly. Brick-earths; but these, if they are used alone, are apt to crack, owing to the quantity of moisture which they contain. This, however, does not hinder persons who understand the business from using them to a good purpose.

"4thly. Strong earths, with a mixture of small gravel, which for that reason cannot serve for making either bricks, tiles or pottery. These gravelly earths are very useful, and the best Pisé is made of them.

"The following appearances indicate that the earth in which they are found is fit for building: when a pickaxe, spade, or plough brings up large lumps of earth at a time; when arable land lies in clods or lumps; when field-mice have made themselves subterraneous passages in the earth; all these are favourable signs. When the roads of a village, having been worn away by the water continually running through them, are lower than the other lands, and the sides of those roads support themselves almost upright, it is a sure mark that Pisé may be executed in that village. One may also discover the fitness of the soil by trying to knead with one's fingers the little clods of earth in the roads, and finding a difficulty in doing it; or by observing the ruts of the road, in which the cart-wheels make a sort of Pisé by their pressure; whenever there are deep ruts on a road, one may be sure of finding abundance of proper earth."

Although a wide range of soils may be used, a suitable relation between the proportion of sand and clay contained in the soil is an essential factor in the choice of the most suitable material with which to build. Typical soils may contain particles of different shapes and sizes, ranging from the rough and angular particles of fine gravel and coarse sand, to the tiny, flat, scale-like particles of clay, and of colloids, which are even smaller. The angular sand particles have only point contact, and there are large voids between them. The minute flat particles of clay, on the other hand, lie together in intimate contact, and there is an infinite number of tiny voids separating them.

When water is added to these two types of soil, the difference in their behaviours is significant. In the case of sand, water passes easily and quickly through the large channels between the particles, and causes very little expansion of the soil. Clay, on the other hand, is relatively impervious, because water is retained in the minute pores by capillary forces. It is, nevertheless, highly porous, and in time each tiny particle becomes surrounded by a film of water and the bulk of the soil is greatly increased. When the water dries out the capillary pressure causes great cohesion of the particles, but extensive shrinkage takes place.

Thus, contrary to popular belief, a soil containing a high proportion of clay is unsuitable for earth building. The high moisture content which is required to make it workable causes extensive surface cracking to take place on drying out, because the inner part of the wall takes longer to dry out, and thus longer to contract, than the outer part. The clay also responds very readily to different atmospheric conditions unless an impervious rendering be provided, and after prolonged repetition of the expansion and contraction of the parts of the wall near the outer surface, without any corresponding movement in the inner part, the face begins to disintegrate.

With sand and sandy gravels, on the other hand, there occurs an almost negligible volume change when the moisture has dried out, and thus there is no tendency towards cracking. This type of soil, however, possesses insufficient binding properties to be used by itself as a building material, and requires the addition of a small proportion of clay to provide the necessary cohesion, and to prevent crumbling.

The most favourable earth mix is one in which the relations are in the region of 25 to 30 per cent. clay, to 75 to 70 per cent. sand. The mix should never contain more than 40 per cent. of clay and colloids, except in the case of blocks, when this limit is not so essential, because most of the severe shrinkage takes place in the smaller mass of each block while curing, before it is laid in the wall.

Silt possesses the defects of both sand and clay, for it is low in cohesion and high in absorption, and a wall constructed of this material would disintegrate more rapidly than one built of clay. Nevertheless, silt has its use in filling the gaps between the larger particles of a soil mix.

The gradation of the aggregate is thus as important in Pisé work as it is in concrete mixes. The aim should be to produce a mixture in which each particle of each different type of material, graded from small stones downwards, is packed round with progressively smaller particles, and which, on compaction, forms a solid mass with the minimum of air voids (Fig. 8). Stones of a larger size than 1 in. in diameter should be discarded, as it will be

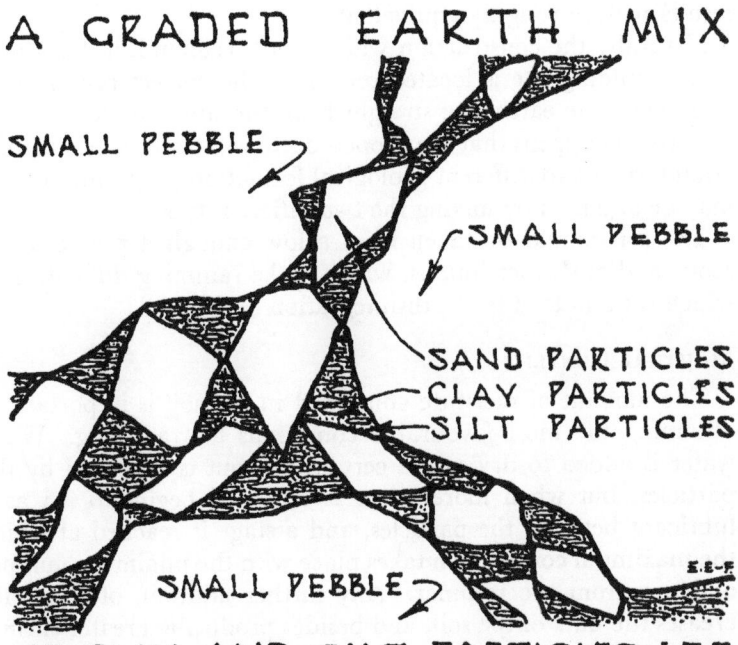

A GRADED EARTH MIX

SMALL PEBBLE

SMALL PEBBLE

SAND PARTICLES
CLAY PARTICLES
SILT PARTICLES

SMALL PEBBLE

N.B. CLAY AND SILT PARTICLES ARE SHOWN LARGER THAN THEY WOULD BE IN PROPORTION TO THE SAND AND PEBBLES

Fig. 8.

47

found impossible to ram them satisfactorily, but pebbles up to this size, if well locked in by the other materials of the mix, will form units of great strength.

A simple analysis of the soil, to show the amount of sand and clay which it contains, may be made as follows: a sample is taken of about three pounds of soil, from which all the large stones have been removed. This is then heated to drive off all the moisture, and is weighed to the nearest ounce. The soil is then put into a shallow pan, covered with water, and gently stirred. When the water has become cloudy with the clay and silt, it is poured away and renewed. This process is repeated until the water will no longer become cloudy, showing that there is no more clay and silt in the mixture. The sample is then dried off again and re-weighed. The result will show the percentage of sand in the soil. This simple test can be carried out more easily by volume, and the results will be roughly equivalent.

Naturally the construction will be most economical if the proposed building site is located on soil of the correct proportions, and where the earth dug straight from the site may be used. It sometimes happens that the proposed site will be situated near the boundary of two different geological formations, and suitable soil may be obtained by mixing the two different types.

The soil should be taken from a low enough depth to avoid roots, and any other humus, which make ramming difficult, and which rot and lead to the disintegration of the wall.

MOISTURE CONTENT

The amount of moisture contained in the soil is important in obtaining the most favourable conditions for ramming. When water is added to dry soil, a certain amount is absorbed by the particles, but when more water is added it begins to act as a lubricant between the particles, and a stage is reached at which the maximum compaction takes place with the minimum number of blows from the rammer. Any further addition of water increases the bulk of the soil, and besides producing greater shrinkage on drying out, also tends to make the soil too plastic and to squeeze out under pressure from the rammer or to adhere to it and be picked up. A soil containing too little water, on the other hand, is found to be too hard and unworkable, and requires a great deal of extra labour in ramming.

A rough guide to the best moisture content is given by taking a handful of the earth, and by squeezing it in the ball of the hand. If the earth just adheres together, it is of the right consistency. A plastic state will indicate an excess of water, and crumbling, too little water.

Another method is to roll out the earth on a flat surface with the palm of the hand until it is about ⅛ in. in diameter. If at this point it breaks in two, it will be found that the moisture content is about right. If it crumbles earlier, it is likely to be too dry, and if it is possible to roll it out still further, it is likely to be too wet. This is the Atterburg "Plastic Limit" test, used by soil scientists, and it so happens that the moisture content at this point corresponds roughly with the optimum for ramming and will be found to be from 10 per cent. by dry weight for a sandy soil to approximately 20 per cent. for a clay type of soil and for chalk.

Before starting to build an actual structure, it is as well to build one or two trial lengths of wall, to find out some of the pitfalls which may be encountered later. When this is done, short lengths of a standard size may be rammed with soil containing different percentages of water, and by giving each a standard number of blows. This method also gives a guide to the optimum moisture content, for it can be seen which piece of test wall is best compacted, and to what extent each has cracked. Care should be taken, when making these trials, to ram the earth in layers of the same depth as will be used on the actual building.

If the earth is too wet, it should be broken up with a spade and left under cover for a day or two to dry. If it is already too dry, water may be sprinkled over it; but it is as well to remember when doing this that it is easy enough to wet the soil, and that it is a more difficult and lengthy process to dry out soil which has been wetted too much.

Were it not for the fact (often somewhat embarrassing) that soil quite incapable of making good Pisé will none the less produce enthusiastic Pisé builders, a warning as to the vital importance of the earth being really suitable might seem superfluous.

The author has found some of the staunchest champions of Pisé-building living on, and valiantly struggling with, stiff glutinous clay and almost pure sand.

Even the most vigorous optimism can achieve little under such adverse conditions unless soil-blending be resorted to, and even so

Pisé-building soon ceases to be economical when complications of this sort are introduced.

Fortunately, however, England is rich in Pisé soils, the red marls being amongst the very best.

A study of the country, or, failing that, of the geological maps, will reveal a great tract of this earth extending diagonally right across England, from Yorkshire down into Devonshire, where it ends conspicuously in the beautiful red cliffs about Torquay. There is a large area of it in the Midlands, notably in Warwickshire, with lesser patches elsewhere in the country.

There is an endless variety of soils that will serve well for Pisé building; some, of course, better than others, but all, save the extremes (the excessively light and the excessively clayey), capable of giving good results under proper treatment.

PISÉ BLOCKS

ANOTHER system of rammed earth building is to use Pisé blocks, which are made by ramming the earth in exactly the same manner as for monolithic walls, except that a small mould is used instead of the shuttering described above for ordinary Pisé work. Pisé blocks have also been made in portable presses.

The blocks may be made in various sizes, an average size being 18 in. by 9 in. by 6 in. A block of this size weighs about 60 lb., which is the near maximum weight that a man can handle. Experiments have been made with the use of thin slabs measuring 18 in. by 18 in. by 3 in., with tongued and grooved edges, and also with types of hollow block; the latter, however, have not been very successful.

It will be found convenient if the blocks are made the full width of the wall, unless the latter is very thick. The block should be formed longer than it is high, in order that it may be laid in the walls with any laminations made during ramming lying horizontally, thus lessening the tendency of the earth to flake away.

A simple wooden press or mould with sides that hinge down is probably the most convenient to use.

The blocks are laid out to dry where they can be protected from the rain, until all the major shrinkage has taken place, and they are then laid in the wall in the same way as bricks, and bonded together with lime or cement-lime mortar, or with a thin mixture of the earth of which the blocks are made. It is important to allow the blocks to dry out as much as possible. A German author states that there can be as much as 5 cm. decrease in the height of a one-storey building as a result of the shrinkage taking place in blocks which are laid too green. Pisé blocks may well be used for partitions and for piers; even chimney stacks may be attempted where the soil is good enough, and where there is a sufficiently skilled workman in constant supervision of the job.

Clearly the advantage of block construction is that it obviates the need for heavy shuttering. It is also easier to detect mistakes in soil mix, moisture content, and ramming, for any cracking and shrinkage will take place before the blocks are laid in the wall.

GENERAL REMARKS

PISÉ is a "dry-earth" method of building, and, as at present practised, that means it is a summer job, so far, at any rate, as England is concerned.

The author is the last person to claim that Pisé building may be successfully and economically carried out in all places, and at all seasons. He merely suggests that in a great many parts of the United Kingdom, Pisé offers possibilities of cheap yet permanent building that are very well worth exploitation.

A wide and thorough trial of the method has been given under a variety of conditions and in a sufficient variety of places. Pisé has been tried out in housing schemes, in Government building programmes, and by ordinary private citizens in need of houses—by the rich (old and new), and by the poor.

If Pisé building is unsuccessfully attempted where the conditions are unsuitable, and in defiance of its physical limitations, the misguided enthusiasts responsible must blame only themselves. But it is not self-reproach alone that they will have to suffer, for the author and all true friends of Pisé will view their troubles with as much anger as sorrow.

Although most of the work of Pisé building consists in digging, carting, and ramming the earth, which can be done by unskilled workmen, the successful execution of the method depends almost completely on the supervision that is given to the job, which must be expert and continuous. A good deal of experience, or at all events intelligent enthusiasm, is required for the managing of the shuttering and the placing of cills, lintols, plates, and, above all, the roof. It is not sufficient that occasional visits to the site be made, as for ordinary brick construction; there must be someone always on the job who understands the work and who takes an intelligent interest in it.

With the exception of the other types of earth walling, there is in all probability no permanent building material for wall construction which is cheaper, provided the building site is located somewhere where earth of a suitable quality is readily available. The capital outlay is particularly low in the case of agricultural building, where spare-time farm labour can be utilized, and where the roof timbers can be made from the farm's stock of wood, and floors of concrete from the local gravel. Pisé has, in fact, been used fairly extensively and with success for the construction of cattle and poultry houses in the United States.

SOME CONTEMPORARY EXAMPLES

ALTHOUGH there is a tendency today to use small percentages of cement with the earth mix to obtain a stronger and more stable material, nevertheless a large number of houses have been designed and built in many countries, within the last two decades, using earth alone as the walling material.

The accompanying illustrations (Figs. 9 and 10) show an elevation and plans, and typical constructional details of designs for some small cottages in Pisé, by Mr. Alex Thorpe, F.R.I.B.A., who is the architect to the Ministry of Agriculture and Fisheries. The arrangement whereby the earth walls are limited to one storey only in height, with the second storey incorporated in the roof, is one which is well suited to Pisé construction. It will also be noticed that the brick chimney stack is carried up in the centre of the house, thus eliminating the necessity of providing a water-tight joint between the earth walling and the brickwork. There are excellent constructional details, such as those preventing the penetration of water at the intersection of the roof with the chimney stack, and also the watertable which is provided to drain the water which drops from the thatched roof well away from the base of the earth walls. In one of the designs, rough, unsawn timber for the roof members is shown, with the idea that it might be procured more easily than sawn timber, from, say, a farm wood store.

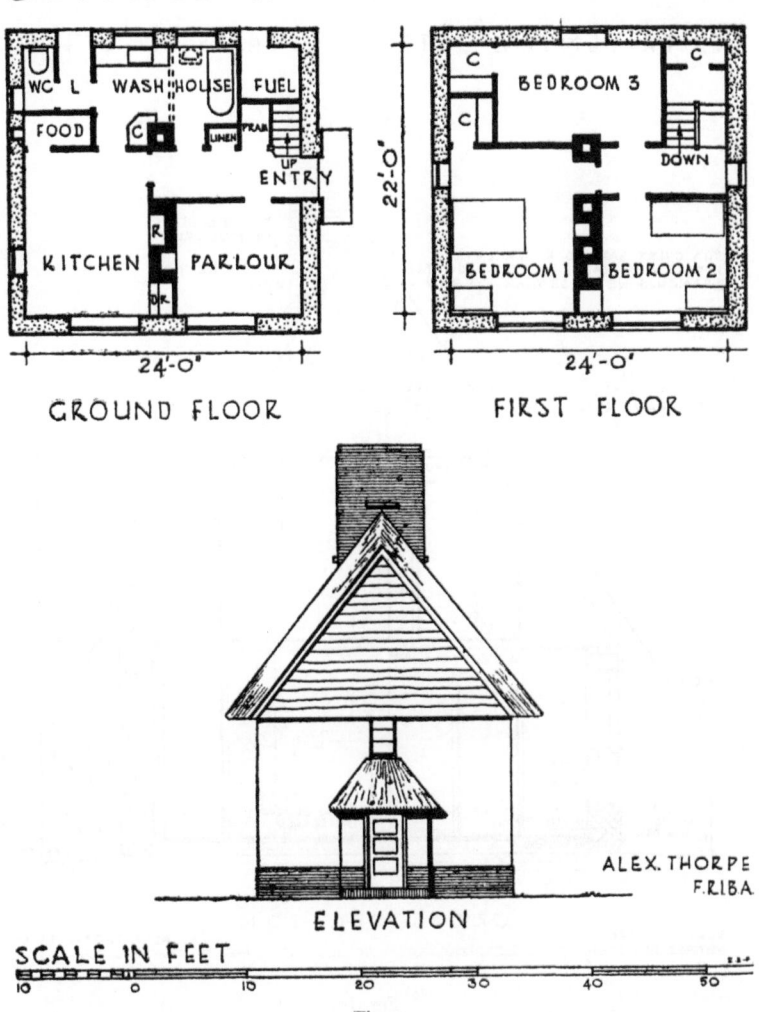

SKETCH FOR PARLOUR
COTTAGE IN PISÉ DE TERRE

GROUND FLOOR

WC L WASH HOUSE FUEL

FOOD C LINEN FRAME

UP ENTRY

KITCHEN R PARLOUR

DR

24'-0"

FIRST FLOOR

C BEDROOM 3 C

C

DOWN

BEDROOM 1 BEDROOM 2

24'-0"

22'-0"

ELEVATION

ALEX. THORPE
F.R.I.B.A.

SCALE IN FEET

10 0 10 20 30 40 50

Fig. 9.

Between 1919 and 1920 the author constructed a small-holder's house at Newlands Corner, near Guildford, to demonstrate the possibilities of using earth as a building material instead of the normal materials which were at that time unprocurable for the building owner of a house of this type. The Mark V shuttering, which is mentioned earlier in this chapter, was used in the construction of the Pisé walls.

EXPERIMENTAL COTTAGES IN CHALK, COB, & THATCH

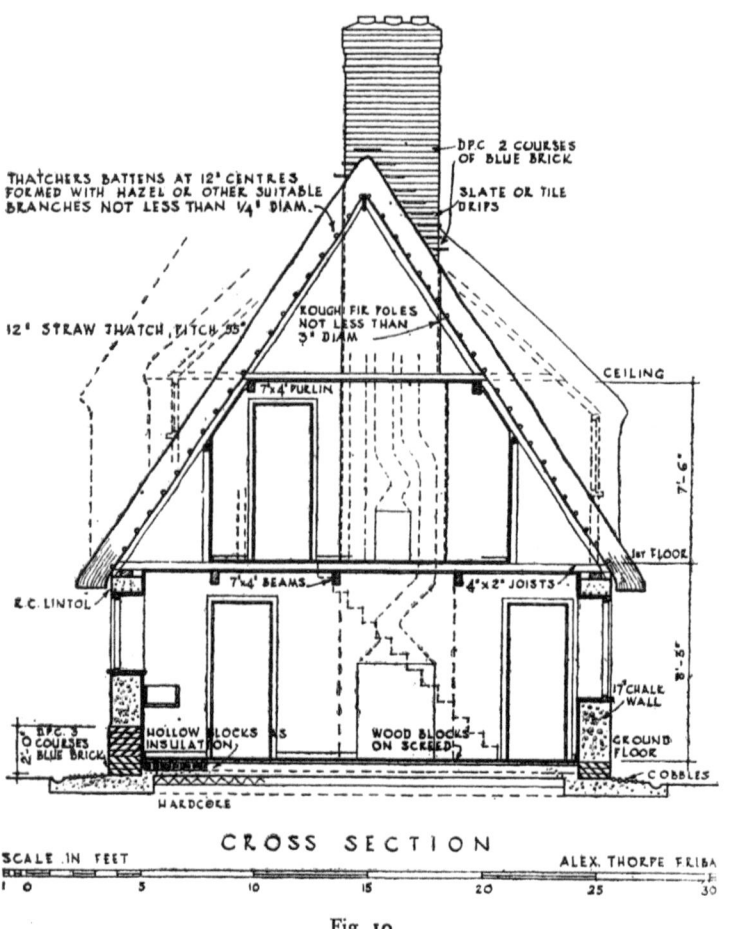

CROSS SECTION

SCALE IN FEET

ALEX. THORPE F.R.I.B.A.

Fig. 10.

The house attracted a good deal of attention from the press, both at home and abroad. It was inspected by multitudes of people, including a great number of Colonials and prospective Colonists, and by many persons directly or indirectly concerned with the problems of housing. The success of the experiment was admitted by all who made the pilgrimage thither. Especially prized amongst the converts was a foreman bricklayer once openly scornful in his unbelief. Of enthusiasm perhaps there was overmuch, and there was some difficulty in restraining the zeal of would-be Pisé builders until the coming of spring, and the return of such weather conditions as the craft might reasonably demand.

The photographs (Pls. 15, 16 & 18) show the general condition of the house in 1945. It is occupied and is still satisfactory, although on closer inspection the protective coverings appear a little the worse for wear owing to lack of maintenance. One desirable feature of design has come to light as a result of the experiment, namely that rain-water pipes should be well blocked out from the walls to protect them from water running down the pipes. The illustration (Pl. 17) shows what may happen when this is not done.

The following is a short description of the house, with an

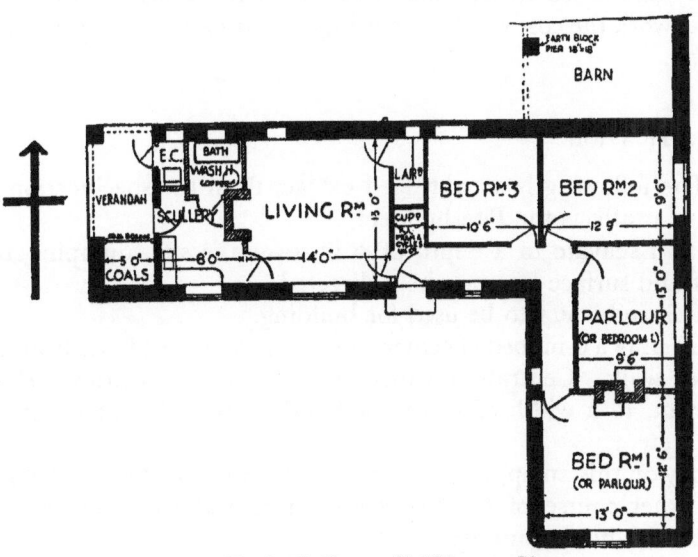

Fig. 11.—Newlands Corner Pisé House. Plan.

abridged extract from the specification containing the information which concerns the Pisé builder, and is published with acknowledgments to the *Spectator*.

The house has six rooms arranged on one floor (Fig. 11).

The walls are of 18-in. solid Pisé work, the roof of red Bridgewater tiles, and the chimney breasts and stacks of brickwork.

The floors are boarded save for the back kitchen, which is tiled. The inner partitions are of 2-in. breeze blocks, the ceilings are plastered, and the casement windows are of steel.

There are two good lofts for storage, one entered from the barn, which is an extension of the house proper.

The pillars of the barn and the partition wall between scullery and veranda are of 18 in. by 9 in. by 9 in. rammed earth blocks; the angle pillar to the veranda is of similar blocks made from soft chalk.

The rest of the structure is of monolithic Pisé, built up *in situ* without joints of any kind, either horizontally or vertically.

Cost

The total cost of the whole of the outer walling of the house (in Pisé) amounted to less than £20. Had the walls been built in brickwork the cost would, according to estimate, have been about £200.

Specification

The following is an abridged extract from the specification so far as it affects the Pisé builder:

(1) Excavate to a depth of 9 in. over the site, dumping the turf and surface humus where directed.

This soil is not to be used for building.

(2) Lay a 6-in. bed of cement and flint concrete 3 ft. wide under outer walls. Centrally on this, lay two courses of brickwork in cement, to a width of 18 in., or build up to the same extent in concrete.

Lay on this an approved damp-proof course; if of slates, having a further course of brickwork or concrete above it to prevent fracture when ramming.

(3) Erect the walls according to the plan on the bases thus formed, carrying them up plumb and true and properly bonded by working round the building course by course, using the special angle-pieces at the corners to keep the work continuous and homogeneous.

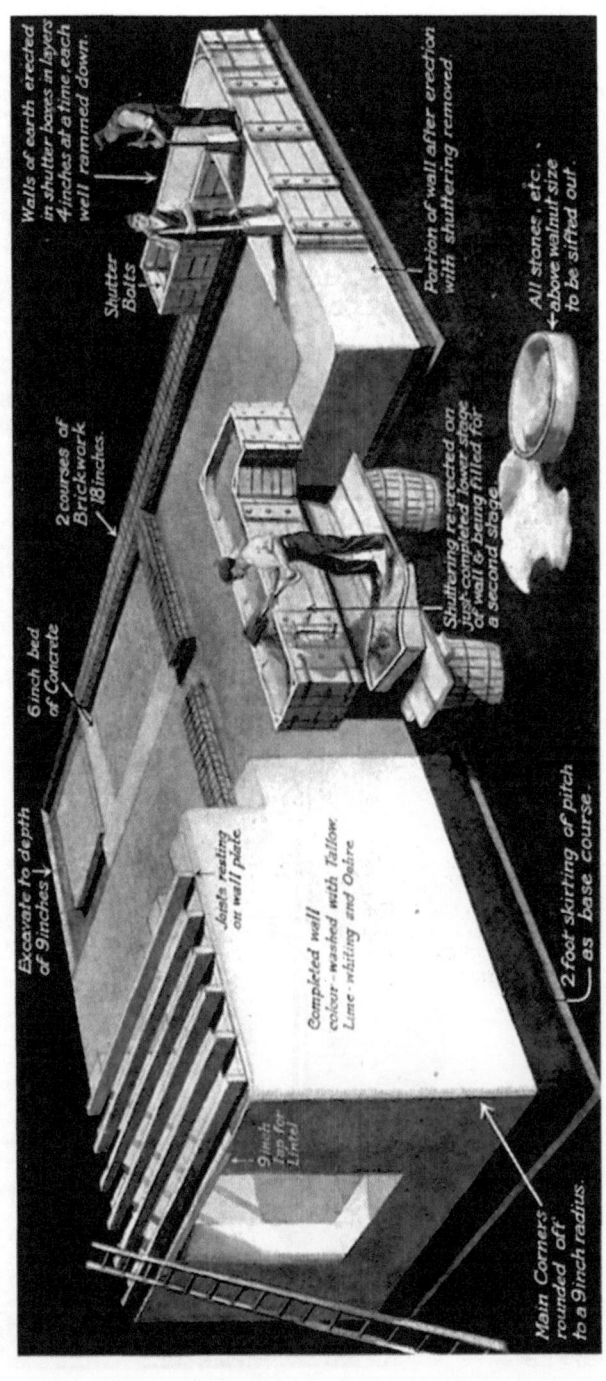

Walls of earth erected in shutter boxes in layers 4 inches at a time, each well rammed down.

Shutter Bolts

Portion of wall after erection with shuttering removed

All stones, etc. above walnut size to be sifted out.

2 courses of Brickwork 18 inches.

6 inch bed of Concrete

Shuttering re-erected on just-completed lower stage of wall & being filled for a second stage

Excavate to depth of 9 inches

Joists resting on wall plate

Completed wall colour-washed with Tallow Lime-whiting and Oubre

2 foot skirting of pitch as base course

9 inch Top for Lintel

Main Corners rounded off to a 9 inch radius.

Pl. 14.—Sketch of a Pisé house in course of erection.

Pl. 15.—The Newlands Corner Pisé Demonstration Building. 1920.

Pl. 17.—One of the findings of the Newlands Experiment— the necessity for blocking rain-water pipes well out from the wall.

Pl. 16.—The Newlands Corner House. The front door, showing the porch formed in the thickness of the wall.

Pl. 18.—The Newlands Corner House. General view showing the condition of the house in 1945.

(4) All stones and flints above a walnut size to be removed by riddling and reserved for concrete.

All sticks, leaves, roots, and other vegetable matter to be eliminated.

(5) The soil immediately on the site to be used without admixture of any sort and to be thrown direct into the shutterings.

No water to be added without the express permission of the architect.

(6) The boxes are to be filled in thin layers of not more than 4 in. at a time, and well rammed until solid. The workmen are not to use their rammers in unison.

(7) Rammed earth at box ends to be shaved down to a 45 degrees slope so as to splice in with new span of Pisé adjoining it.

Where door and window openings occur, the special "stops" to be adjusted and firmly secured so as to withstand hard ramming. Two 4-in. by 2-in. by 9-in. plugs to be built in to each window jamb for the securing of the frames and three to each door jamb.

Special care to be taken in the thorough ramming at the corners and along the box edges.

(8) Insert below floor level, where directed, 24 3-in. field drainage pipes to act as ventilators through the thickness of the wall. Insert wire-mesh stops to exclude vermin.

(9) Set all frames square and plumb, and where in outer walls, flush with finished exterior plaster-face, the joint being covered by a 2-in. by $\frac{3}{4}$-in. fillet.

Where lintols occur, they are to be tailed in at least 9 in. on each side of the opening.

Provide plain picture-rail round all rooms at window-head level, providing plugs for fixing where necessary.

Secure to floor round all boarded rooms a 2-in. by $1\frac{1}{2}$-in. angle fillet as skirting.

(10) The smooth surface of the Pisé walling to be hammer-chipped to give good key to the plaster.

Before rendering or plastering walls, any loose earth or dust to be removed with a stiff brush and the wall surface evenly wetted.

The rendering to be carried evenly round the walls—the minor square angles being roughly chipped down first so as to obviate sharp corners. The main corners of the house are ready-rounded off to a 9-in. radius by the special corner mould.

(11) Bond brick and slab work to Pisé walls by driving iron spikes into the latter every few courses at joint level and bedding in.

(12) Colour-wash walls with tallow lime-whiting tinted with ochre. Provide 2 ft. skirting of pitch, applied hot, to form base-course round exterior of building.

CHAPTER THREE

Adobe

Adobe, which is the Spanish word for "mud," is the name given to a certain type of block construction in which the earth mix is similar to that of cob and clay lump. It is common in Mexico, especially around Las Cruces in the north, and is often practised in California and the south-west of America, where other materials for building are scarce and where it has great durability as a result of the semi-arid climate. Many of the old Missions and homes in California were built of it, and it is reported that in that State an old Adobe block house stands at Monterey, which was built in the early nineteenth century, and which has been occupied continuously as a dwelling-house for over one hundred years. Doubtless there are many older houses of the same type, their adobe walls lying concealed beneath various kinds of rendering.

The following extract from *The Farmer's Handbook*, issued by the Department of Agriculture, New South Wales, 1911, will serve as an introduction to the method.

"As their name (Adobe) implies, these buildings are constructed of sun-dried, but unburnt bricks. For buildings of this character, material like clay, which is unsuitable for Pisé-work, can be used. The bricks are made in a wooden mould, and are 16 in. long, 8 in. wide, and 6 in. thick. A man can mould about 100 per day. They are laid in a similar manner to other bricks, the mortar used being wet loam, or even the material of which the bricks are made. The cost of making and laying is estimated at about 15*s*. per 100. Buildings constructed of these bricks are substantial and cool, and very similar in character to Pisé buildings.

"A school-house built of these bricks eighteen years ago by Mr. Nixon, of Reefton, is still in an excellent state of preservation; in fact, little, if any, the worse for wear, despite the fact that walls are unprotected by verandahs or overhanging eaves. During its existence it has had, first one coat of oil-paint, and later a coat of coloured limewash."

SOIL PREPARATION

The soil for making the blocks is mixed up by hand in a puddle or in a depression in the ground, water being added to it and mixed in with the feet or with a farm implement until the whole is turned into a sticky mass which can be picked up with a fork. The preliminary mixing of the earth can conveniently be done by driving a disc-harrow back and forth over the soil heap.

MOULDING

When the soil is thoroughly mixed, it is placed in the moulds. These moulds, which have neither top nor bottom, are laid lengthwise on the bare ground or upon a thin layer of straw, and the soil

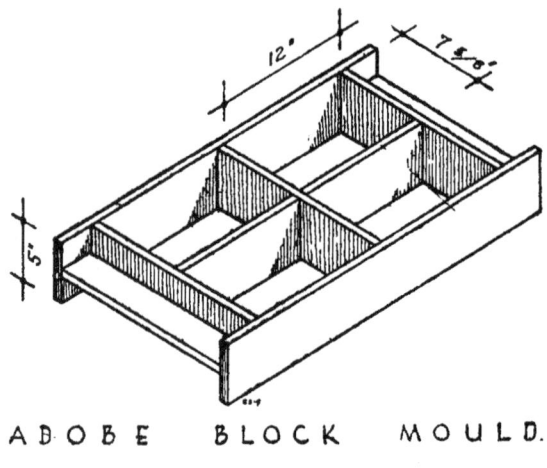

ADOBE BLOCK MOULD.

Fig. 12.

is pressed in with the feet, and with a flat tamper to ensure that the corners of the mould are properly filled. The excess earth is cut off flush with the top of the mould. Often the mould is so constructed that more than one brick can be made at once (Fig. 12).

A sheet-metal lining to the moulds helps to stop the mud from adhering to them, but is not essential.

CURING

After moulding, the blocks are turned out, and during this process they become slightly concave because the edges adhere to the mould. They are left out to dry for two or three days, care being taken that the rain does not get at them. After this preliminary drying, the blocks are turned up on edge so that air can circulate round the parts which are still damp; they usually dry in a week or so according to the weather. They are then stacked in such a way that they are tilted over at about 45 degrees, so that they are resting on the sharp edge along one side, allowing air to circulate freely round the greater part of all the sides. The stack must be protected from the rain.

Size of Block

When a sufficient number of bricks has been made the construction of the building is begun. The size of the blocks is determined by the desired wall-thickness, and by the weight which can be handled conveniently; the latter is found to be not more than sixty pounds, and the average size of an Adobe block is usually about 18 in. by 12 in. by 4 in., walls of one storey being commonly built 12 in. thick.

Laying

The blocks are laid in the same way as ordinary bricks, and the joints are made with a thin mixture of the earth used for the blocks, or with a weak lime mortar. The use of a cement mortar is not to be recommended, since it will set stronger than the blocks, and cracks at the joints are likely to result. Earth is said to make a good mortar in a warm, even-temperatured climate, but in those countries where the temperature fluctuates to greater extremes it is thought that a rather stronger mortar may be necessary. The mortar joints should be just thick enough to true up the course. Galvanized wire is sometimes incorporated in them, and is afterwards bent over to hold chicken wire which serves as a key for rendering. For although bare Adobe walls are known to remain in good condition in a dry climate, some form of external rendering is necessary to protect them if they are required to endure for a long time, especially in wetter climates. Different types of rendering are discussed in Chapter VI.

Soil Mix

The soil mix for Adobe blocks does not have to be so carefully proportioned as that required for Pisé work, since the most serious cracking and shrinkage can take place before the blocks are laid in the wall. It is, however, generally recognized that the soil should contain at least 50 per cent. of sand to make a block which is sufficiently strong and stable. If the soil contains too great a proportion of clay, it is often reinforced by the inclusion of some binding material such as straw or grass fibres. Some authorities say that this fibrous material provides a number of small cleavage planes along which shrinkage cracking may take place, instead of occurring more seriously in fewer places. An excess of sand will produce a block which is too friable, and it is well for the intending builder to make a few test blocks before embarking upon the construction of the actual building.

WATER CONTENT

Water should be added to the mix until the soil is of a suitable consistency for moulding the blocks. At this stage the soil will be conveniently plastic, but capable of holding its own shape when the mould is removed. If the mix be too wet the block will slump. The United States Bureau of Standards [1] quotes a figure of 16 to 20 per cent. of water by weight compared with the dry weight of the material, as likely to give the required consistency. A mixture containing less than 16 per cent. of moisture, it is stated, will be too stiff and will adhere to the mould, making the block very concave and thus cracking the lower surface. Over 20 per cent. moisture will cause the earth to slump and lose its shape upon removal of the mould, and shrinkage cracks will appear on drying out. This figure is considerably higher than that required for Pisé work.

UNBURNED CLAY AND EARTH BRICKS IN ENGLAND

The use of sun-dried bricks in this country is, for no very apparent reason, almost entirely restricted to East Anglia. There it has been used for generations with entirely satisfactory results.

Mr. Skipper of Norwich wrote of the material as follows:

"Who, travelling from Norfolk to London, whether by the Ipswich or Cambridge line, has not noticed the numerous colour-washed or black (tarred) cottage, farmhouse and agricultural

[1] U.S. Department of Commerce, National Bureau of Standards Report, B.M.S. 78. *Structural, Heat Transfer, and Water Permeability Properties of Five Earth-wall Constructions.*

buildings scattered practically all along the countryside? Some
of these are of studwork and plaster, some of wattle and daub, but
many are built of clay made up into lumps, sun-dried, and built
into the walls with a soft clay-mixture as mortar. No lime *need* be
used, though sometimes it is mixed with the clay mortar. The
preparation, digging, exposure, and mixing with short straw are
similar to the Devonshire 'cob' work, but in these parts the
worked clay is thrown into moulds, and lumps are formed of, say,
18 in. by 12 in. by 6 in., or 18 in. by 9 in. by 6 in. for large sizes,
and for inside walling or backing to brick-faced walls, 18 in. by
6 in. by 6 in. The walls, naturally, are rough in texture and the
joints are generally stopped up and besmeared with a thin coating
or almost a wash of clay. This coating sometimes has lime mixed
with it, but it is not necessary. This is all that is needed to complete the walling, and there is a building—a malting, that any one
can see at Tivetshall Station on the Ipswich line, about 200 ft.
long, 45 ft. or 50 ft. wide and three floors high, built of lumps
18 in. by 12 in. by 6 in.—that has stood the weather and weight
of its roof for forty years built in this way; 12 in. is the thickness of
its walls. A further stage in finish is to give the walls two or three
coats of coal tar, but it is not essential, though desirable where
stock are kept, as cattle are rather fond of licking the clay, and
they do not use their horns much when walls are tarred. The
highest finish in this work is to cast sand on the last coating of tar
before it is quite dry, and then to colour or whitewash on this.
This accounts for the variety of colourings seen in these buildings,
some even of a kind of pink or red; while some yellow or buff,
beside the white and the black or tarred buildings, and all
huddled together or standing apart, whether covered with thatch
or red pan or flat tiles, look remarkably in harmony with their
surroundings. These lump walls are, of course, built on a base of
brickwork, about 18 in. or 2 ft. high, to keep them free from damp.
This kind of walling can be built for *at least* 15 per cent. or 20 per
cent. cheaper than ordinary 9-in. brickwork. Thin as these walls
are compared with those of 'cob' houses, they are noted for being
warm in winter and cool in summer. When suitable clay is pro-
curable a local builder almost invariably uses clay lumps when
building a house for himself, though to gratify a whim perhaps, he
will case the outside walls—especially the front next the street or
road—with brickwork. But clay lumps he carefully reserves for

inside walls and weight-carrying linings to the outside walls, bonding the two together very much in the same way as two $4\frac{1}{2}$- in. 'cavity walls' are bonded. I am not suggesting that this walling is as interesting artistically as 'cob,' but I do suggest it is a practical, sensible and *dry* walling, and if properly done it will 'last for ever,' as a local builder repeatedly said to me when speaking of it. One can easily see why the cost is light—the sun and the winds do the drying in the spring months, and no coals are required, and also the clay is often found on the building site, hence no cartage. Actual building work naturally goes quickly, as the lumps are large. There is another important point to notice. One may see a building complete with its roof on and occupied by its tenant while still awaiting an outside casing of brickwork to be built round it, either with a view to greater protection or for the mere vanity of the owner, for while thus left unprotected the lump walls take no harm from even winter exposure. Now to be quite practical in these extremely practical days, I venture to suggest that the use of clay lumps at least for inside walls and linings of outside walls would be an immense boon to the numerous cottage-building schemes now being projected. We must not forget that comparatively few bricks will be available this year, while the cottages are wanted at once. Can these few bricks be better used than by forming foundations and chimneys for the clay-lump walls of these cottages? I think not. The cottages could, of course, be occupied in the late summer or autumn of this year, and next year when bricks will be more plentiful perhaps the brick casings could be added, if brickwork *must* complete them. I make this strictly utilitarian suggestion solely to meet a very urgent and deep national need. Personally, I prefer the sight of a cottage built and finished in the old-established method of the locality. Unskilled labour only is required, working under intelligent supervision, hence immediate employment for a great number of men would be provided."

The use of sun-dried bricks for the interior partitions of Cob and Pisé cottages is worth consideration, as the nature of these materials demands a thickness of wall which is too wasteful of space to be acceptable in mere partitioning.

Of the strength of clay-lump walls, there is no question. At the time of the first edition of this book in 1919, it was necessary to cut a new doorway in the old clay-lump wall of a large traction-

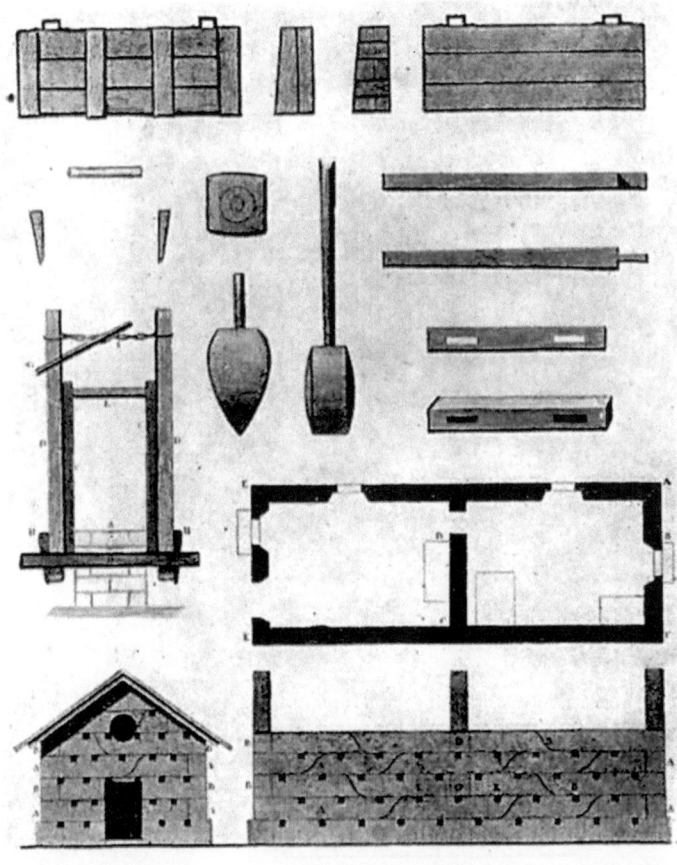

Pl. 19.—Pisé Plant and Implements. Reproduced from an old Encyclopædia.

Pl. 20.—Engineering Workshops built about forty years ago. The walls were thoroughly sound, despite constant vibration, and perfectly dry when the photograph was taken twenty years ago.

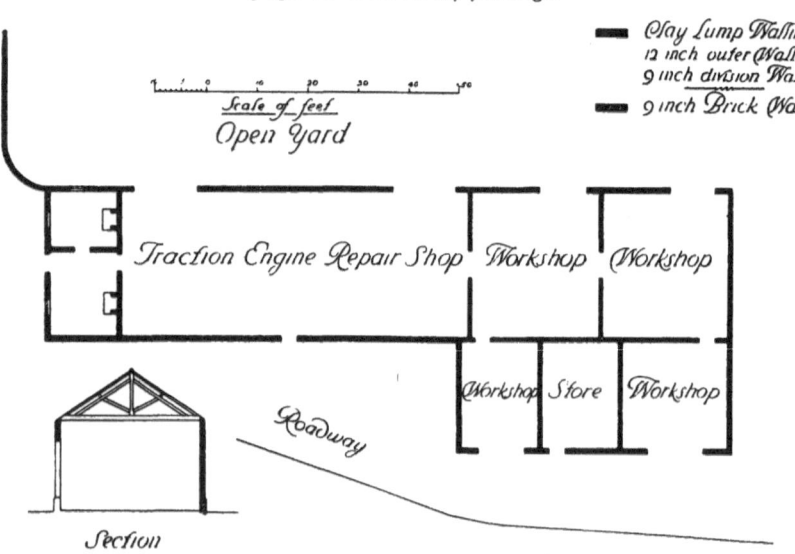

Clay Lump Walling
12 inch outer Walls
9 inch division Walls
9 inch Brick Wall

Scale of feet
Open Yard

Traction Engine Repair Shop   Workshop   Workshop

Workshop   Store   Workshop

Roadway

Section

Pl. 21.—Plan of Engineering Workshops.

Pl. 22.—After one hundred years' service. This Council School, once a Corn Hall, was built of Clay Lump.

Pl. 23.—A Row of Clay-lump Cottages. The front has been plastered and panelled out. In the upper part of the stable building, seen in the foreground, the Clay Lumps are shown exposed.

Pl. 24.—Ramming Terracrete wall for tests at National Bureau of Standards, Washington, U.S.A.

Pl. 25.—Terracrete house, stuccoed, in Orange, Virginia, U.S.A.

ADOBE

engine garage, and the blocks removed were thrown into a heap upon the ground.

The clay happened to be needed for other purposes, for which it had first to be broken up.

Ordinary hammers proved entirely ineffective, and it was not until heavy sledges were used that the lumps could be smashed.

The tractor-house in question was a large building some 25 ft. by 100 ft., carrying a heavy roof and constantly subjected to vibration by the coming and going of the tractors.

The walls were only 12 in. thick, without piers or reinforcements of any kind, and yet the whole building, which was 26 ft. high at the gables, was as perfect then as when first erected some twenty years previously.

In the same town as this tractor-house, East Harling in Norfolk, was a council school built of clay lump (converted from the old Corn Hall), apparently not a pin the worse for a century of hard wear (Pl. 22).

Near by there were a number of private houses built of the same material, some of them reputed to be upwards of 200 years old and certain of them having considerable architectural merit.

CHAPTER FOUR

Stabilized Earth

BROADLY, "Stabilization" can be thought of as any process by which earth is made less liable to volume change, more resistant to water, and of greater strength and hardness. Some soils are naturally stable, and those containing a high proportion of well-graded sand and small gravel and sufficient clay and colloids to bind the whole together, but not to cause undue shrinkage cracking on drying out; are the most stable and the most suitable for earth walls. Some of these naturally occurring deposits are known as "hoggin," and they exist in nature, moreover, at their optimum moisture content for ramming.

The traditional earth wall was relatively stable because roughly just such a graded mix was used, and because the best protection possible was applied to keep it dry. In practice today, however, "Stabilization" usually implies the addition of a special "stabilizer" to the earth mix.

Although the study of soil mechanics is as yet a relatively young science, considerable attention has already been given to the subject of soil stabilization for various purposes.

The stabilizing agents fall into two classes: those which impede moisture penetration by blocking the capillary pores, and which may or may not add materially to the strength of the wall; and those such as resins and waxy oils which are thought to prevent capillary rise by providing a thin water-repellent skin on the surface of the film of water surrounding each earth particle, but which have no strength of themselves and maintain the strength of the soil only at its naturally highest level. The various stabilizers are discussed later under their separate headings.

In the early thirties some of the first scientific work was carried out, principally in the United States; more recently the subject has been studied in many parts of the world, and particularly in relation to the programme of airfield construction which has taken place during the war.

One of the results of all this research, particularly, again, in America, has been the application of the principles of stabilization to earth building. Research into this special field has, in fact, been carried out at a number of places, among the most important being the University of Illinois, the South Dakota State College, and the U.S. Bureau of Standards, where many laboratory analyses and actual weathering tests have been made (Pl. 24).

HISTORICAL BACKGROUND

There is little to relate of the historical background to Stabilized Earth building, since interest in the technique has only recently developed. Nevertheless, there were a few experiments carried out at an early date which are interesting to note, as showing the way in which the problem was first approached.

After the war of 1914–18, when the greater part of Ypres had been razed to the ground, an acute housing shortage, linked with extreme difficulties of transport, gave rise to the erection of some experimental houses in which a form of stabilized soil was used.[1] The houses were erected by Messrs. Holland, Hannen & Cubitt, and were built of brick earth and debris found near the site. This material was spread out and screened, $\frac{1}{10}$ to $\frac{1}{6}$ of its volume of pulverized hydraulic lime was added, together with chemicals for preventing organic growth, and the mix was again screened and then rammed. The walls were made impermeable by a coating of a patent liquid containing two parts crude benzol and one part bitumen previously dissolved in benzoline, and resin and quicklime, which was applied to the exterior face, penetrating well and having a good hardening effect. The construction consisted of a concrete plinth wall with damp-proof course upon which were constructed reinforced concrete piers, having horizontal string courses, also in concrete, running between them at first-floor level in those houses which had two storeys. The earth was used for the panels between the concrete piers, and, after the bitumen solution had been applied, was painted with a mixture of lithopone, benzoline, and resin. More houses were built later in other parts of Belgium by this method, which was called "La Terrademente."

Hydraulic lime was used for stabilizing soil also, by Mr.

[1] Building Research Station. *Special Report, No. 5.* P. W. Barnett and others. H.M.S.O., 1922.

COB, PISÉ, AND STABILIZED EARTH

A. H. H. Scott,[1] who constructed a number of test walls at about the same time as the Ypres houses were being built. Among the test walls was one of chalk and blue lias lime, in the proportions 10 parts of chalk to one of blue lias lime, to which a gallon of integral water-proofing mixture was added. This mix was said to give results next in order of soundness to Portland cement mixes.

In 1920, the Department of Scientific and Industrial Research carried out their experiment in earth building at Amesbury. In one of the cottages which they built, a small proportion of cement was added to a chalky earth which was then rammed as in Pisé construction. The cottage was still in very good condition in 1945, and an account of the mixes used and the method of building is given in a later chapter which is devoted entirely to this important experiment (see Chapter VIII).

CEMENT STABILIZED EARTH

ONLY small percentages of dry cement need be added to a soil mix to make it less vulnerable to the effects of water, and have considerably greater strength than it would otherwise have. In America rammed earth, stabilized in this way, has been given the special name "Terracrete" by Mr. Francis Macdonald, and his booklet on the subject has that title[2] (Pl. 25).

PROPORTIONS

With good Pisé soils, that is, soils of a predominantly sandy type and containing only about 25 per cent. to 30 per cent. of fine material, between 4 per cent. and 7 per cent. of dry cement by weight is required in the mix to give a marked hardening effect to the soil. The more clayey the soil, the more naturally unstable it will be, and consequently the more cement will be required to give a satisfactory stabilizing and hardening effect. The clayey soils may be substantially hardened by the addition of between 6 per cent. and 10 per cent. of dry cement. However, the use of such a large proportion of cement will not in all cases justify itself, on grounds of expense, and the transporting of sand to the site for

[1] P. W. Barnett, *loc. cit.*
[2] F. Macdonald, *Terracrete—Building with Rammed Earth Cement.* Chestertown, Maryland, U.S.A., 1939.

mixing with the local soil so that it is possible to use a lower percentage of cement to achieve stability may prove to be a cheaper proposition. As digging is involved, whether the soil be taken from the building site or from elsewhere, and constitutes a large proportion of the total labour in building, the extra expense of importing better soil consists only of the actual price of transport, plus the cost of the soil, if it has to be paid for. A stabilized heavy clay will not in many cases possess any greater strength and durability than a well-proportioned plain Pisé mix, for the earth will be difficult to mix and the cement will not penetrate the unbroken clods which will constitute units of weakness.

Moisture Content

When cement is used, it is even more important to obtain the correct moisture content for the mix than with ordinary Pisé work, because shrinkage cracking may be correspondingly more severe. This was demonstrated in the Amesbury experiment in the cottage known as No. 10 Holder's Road (see Chapter VIII). The shrinkage cracking which occurred in the cement stabilized walls was found to be much more serious than that which occurred in those of a similar cottage of Pisé.

As a result of researches into soil-cement mixes for light road construction, the U.S. Bureau of Standards gives a figure of about 10 per cent. to 12 per cent. for the best moisture content for ramming. The amount of water required for the hydration of the cement can be neglected in estimating the optimum moisture content, since it can be assumed that hydration does not take place until after the soil has been rammed, and that the cement does not therefore, initially, take up any of the moisture from the soil which is required for ramming.

Mixing

A rough method of mixing which has proved satisfactory where it is not necessary to ensure great strength in the wall is to spread the cement on the ground straight from the bags, and to drive a disc-harrow to and fro until the cement is well mixed in with the soil (Pl. 26).

The use of a concrete mixer will, however, give more accurate mixing. The process is as follows: a batch of soil is dried on a wooden platform, placed in the concrete-mixer, and turned over and over until all the lumps are broken up; the cement is then added and mixed until it is well distributed and the mix has an even-coloured appearance. Lastly, the required amount of water is added, and after being well mixed in, the soil-cement is rammed into the moulds within two hours of mixing, before setting takes place.

CURING

The completed sections of walling should be kept damp, either by spraying or by covering with sacking, and should, at the same time, be protected from the hot sun, for at least three to seven days, and preferably longer, until the cement has been cured. The tops of the walls should also be protected from rain when they are quite green, but since they harden-up quickly there is not the same need to keep them covered until the next course is rammed, as with ordinary Pisé.

PLATING

The techniques of Pisé and Stabilized Earth are sometimes combined in a system known as "Plating." The rammed earth wall is given a facing of soil-cement, about 3 in. deep on the exterior side of the wall, to give added protection against the weather. The soil-cement is often carried in a layer across the whole thickness of the wall above the lintols for extra strength, and below the window openings to form internal cills. It can also be used with the Pisé in the form of studs of the same thickness as the wall, between which panel infillings of plain earth faced with soil-cement are rammed.

The procedure for plating is simple: the soil-cement is placed against the outward side of the shutter and the rest of the shutter is filled with the ordinary Pisé earth. The whole is then rammed in the ordinary manner for Pisé work. It is not important that the plating should be exactly 3 in. thick all over the walls, as this is too difficult to achieve; it is sufficient that a thick skin of soil-cement be maintained over the wall face which will blend gradually into the common Pisé behind (see Fig. 13). Plating a poor Pisé soil with a good soil-cement mix which has a high sand content is not recommended, because the two soils will not mix together successfully, and the plating will, most likely, fail to adhere to the walls.

Soil-cement may often be used with advantage for piers or studs (Fig. 14), in conjunction with an ordinary Pisé or a plated building.

The interest which the Germans have taken in the use of earth for building is discussed below (Chapter IX), but it is interesting

PLATING WITH SOIL-CEMENT

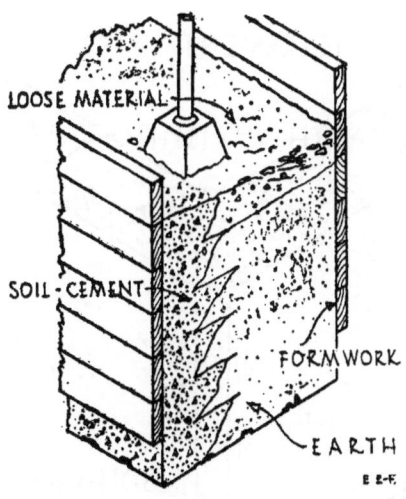

Fig. 13.

to note here that a German writer [1] has advocated the use of plating to improve earth walls against penetration by water, and to minimize their tendency to dusting. The plating is worked into the outside corner of the form to a thickness of 3 to 5 cm., either with a trowel or by means of a straight-edge. After the earth has been added the whole layer is then rammed in the usual way. The plating which is recommended is one made of cement, earth, and gravel, composed of stones of from 1 to 3 cm. in size.

In the same article another method of plating, invented by one Baurat Fauth, is described. Chaff, of about 5 cm. in length, is mixed up thoroughly with mud so that each of the pieces is well coated. It is then stored damp until it is required. Cement

[1] *Bauwelt*, February 15th, 1944. Number 3/4.

mortar is then added wet in the proportions of 1 to 6, and is mixed in well, so that each piece of the muddy chaff is covered with the mortar. The mixture is placed in the shuttering with a trowel, is rammed as before, and is said to give added weather protection to earth of a poor quality, as well as a good key for rendering if

PLATING WITH EARTH INFILLING

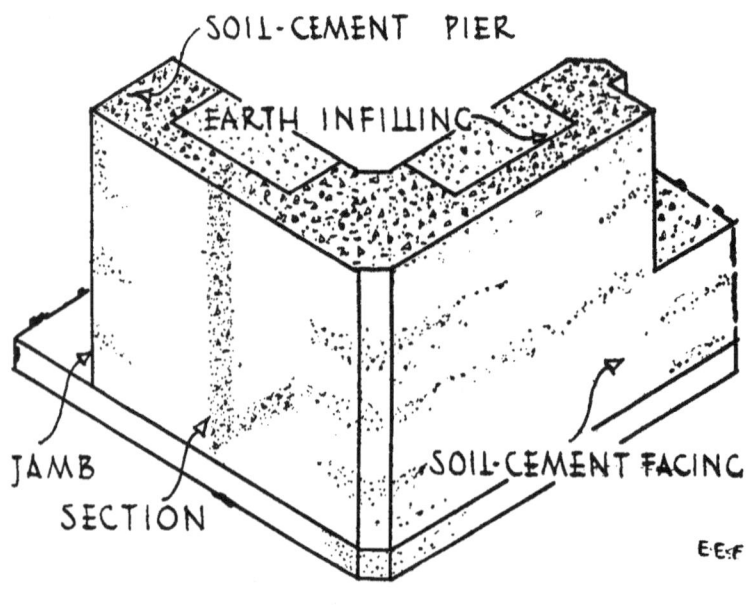

Fig. 14.

required. It is also said to make a good bond between one layer and the next. Warning is given against trying to ram a rendering in the shuttering in the same way as plating, for if this be done the rendering may crack, since it will dry out and shrink much faster than the walls.

For permanent buildings the plated wall should be covered with some form of rendering to make absolutely certain that it is quite invulnerable to water; but in temporary buildings, such as outhouses, garages, and other buildings where long years of service are not required, the soil-cement plating may be left exposed.

It should be remembered that plated walls need to be kept damp for several days in the same way as those of solid soil-cement, in order that they may be thoroughly cured.

Costs

The exact costs of the various items in earth building cannot be given, as they depend to such a great extent upon the local factors influencing the construction, such as the availability and type of labour, and of soil. Thus, in the case of cheap agricultural buildings, it is not possible to say whether a system of plating would prove to be a cheaper alternative to solid Pisé walls protected by an impervious rendering. Nor is it generally possible to say whether the use of plating would be any more economical than the use of solid soil-cement walls, because in the latter case the thickness of the walls could, for a single-storey building, be reduced from about 15 in. to 12 in. The disadvantage of this, however, is that the reduction of wall thickness causes also a reduction in thermal insulation. The most suitable method must be decided upon according to the use to which the building is to be put, and after consideration of the local conditions already discussed.

Shuttering

Soil-cement is more liable to stick to the shutters than is ordinary Pisé soil, and in order to overcome this, an application of oil such as used crank-case oil should be given to the inner face of the shutters after every few lifts. With very sandy soils frequent oiling is necessary even with common Pisé.

Supervision

When supervising the work on site it is most important that the following points should be kept in mind. In the first place, despite being given full instructions as to the proportion of earth to cement, the workmen tend nearly always to insist on using more cement than is required, believing that it will make the mix stronger, and little realizing the disastrous effects of using such a rich mix. Secondly, the need for adequate mixing must be impressed upon the workman. The lumps of earth must be broken up, and since the cement forms such a small proportion of the whole mix, it must be well dispersed. The third factor is that the correct moisture content plays a most important part in successful building, and, although it is better to err on the side of a mix that is slightly too wet, it is only too easy for the measuring of the water to become a little careless, and for there to be a strong tendency to make the soil very soft and plastic by adding too much water, so inviting cracks on drying out.

KEY FOR RENDERINGS

As renderings will not adhere to the relatively smooth face of earth walling, some form of key must be provided. A German practice in Pisé work is to place small flat fragments of tile into

KEY FOR RENDERINGS

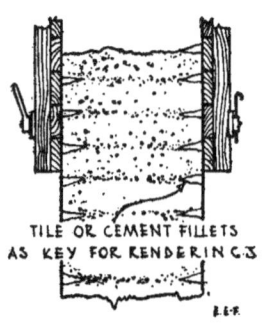

TILE OR CEMENT FILLETS
AS KEY FOR RENDERINGS

Fig. 15.

each layer as it is rammed.[1] The earth immediately above the fragments is scraped away as soon as the shuttering is removed, to provide the key. This method works well, apparently, but it is not always possible to procure the tile fragments. Stabilized Earth can be used instead, and fillets should then be made of a gravelly earth, with cement or hydraulic lime, in a dry mix of the proportions 1 : 4 or 1 : 6; and only enough of the material should be made up for immediate use (see Fig. 15). The coarse mortar should be applied with a trowel along the edge of each layer of earth, against the shuttering, and the next layer of earth should then be

[1] *Bauwelt, loc. cit.*

74

rammed on top of it. The procedure should be repeated with each layer of earth in the wall. The fillets need not be large, but should be placed on both sides of the wall in order to avoid unequal settlement. It is claimed that a very good key is provided if the mix contains enough coarse material of about 1 to 2 cm. in size; the layers of stabilized fillets are also a rough guide as to whether the courses are being laid horizontal.

DESIGN

The principles of design for soil-cement buildings are much the same as those for buildings in other types of earth walling (see Chapter VII). However, since a good soil-cement is harder than rammed earth, and should stand a test of three days' immersion in water without showing signs of disintegration, the concrete or masonry plinth wall, for instance, need not necessarily be carried up so far as in Pisé. It should be remembered, however, that some types of damp-proof course are easily fractured and punctured, and that if, as a result of the plinth wall being lower, they are placed on top of it instead of being inserted in the middle of it, they should be protected by a further layer of concrete or masonry.

Soil-cement is said to be immune from attack by rodents, and by termites in tropical countries, so the precautionary measures taken against them in building the lowest course of a Pisé wall should not be necessary in a Stabilized Earth wall, provided the mix contains sufficient cement.

Likewise, precautions taken in a Pisé and Cob building to protect the walling against water are also not generally so vital to a building constructed of soil-cement; but in order to obtain a thoroughly durable wall, it is worth while observing the rules of design which are common to the other types of earth walling and to give as much protection as possible.

Another feature of some rammed soils stabilized with cement is that nails may be driven into them, and will hold well without causing the surface of the wall to flake away. Nevertheless, fixing blocks will inevitably be required for some purposes and, after they have been treated with preservative, they should be incorporated in the walls as they are rammed. Mr. Macdonald makes particular mention in his booklet of the tendency for the blocks to dance and become displaced when a pneumatic hammer is used for ramming, and in order to avoid this he suggests that the earth immediately surrounding the blocks should be rammed with a hand rammer and that a strip of metal lathing be fixed to the underside of the block, projecting several inches at each end, to help keep it in place.

It is usually impossible to remove the shuttering, and to carry out subsequent building operations, without the walls suffering minor damage. It frequently happens, for instance, that when the shutters are removed pockets remain in the wall surface where the earth has not been rammed up close to the sides of the shutters, and if it is not intended to render the walls, they may be made good by a trowelled application of mortar, of a slightly wetter mix of the soil and cement used for the walls.

During 1944–45 some work was done by The Cement and Concrete Association to ascertain the suitability of soil-cement for farm buildings, and to experiment with methods of mixing. In constructing a number of test walls, some samples were given a smooth finish, by trowelling the surface of the newly completed wall as soon as the formwork had been removed (Pl. 27). Irregularities in walls could be eliminated in this way and the smooth surface obtained might prove useful for internal finishes for houses; for instance, slurries could be applied instead of plastering, or the walls might be papered direct.

SOIL-CEMENT BLOCKS

SEPARATE blocks may be made of soil-cement in the same way as Pisé blocks are made. They should be cured for at least 10 days before being laid in the wall, when they are less likely to give trouble from shrinkage cracking than a monolithic wall. The mortar may be composed of the same material as the blocks are made of, or it may be a lime or cement-lime mortar. There is a tendency for these blocks to become somewhat friable, and they should be handled carefully, but they are useful not only for the main walling of a building, but also for gable ends, partitions, fireplaces, and elsewhere where shuttering would be awkward to handle. Ordinary shuttering may be made into a mould for blocks by cleating a number of stop-ends into it at suitable distances from one another to make the required size of block.

STABILIZED EARTH

POSTSCRIPT

As a postscript to this discussion on cement Stabilized Earth, the following example has been worked out to give a rough indication of the approximate amount of cement which might be required (exclusive of foundations) in the construction of a small bungalow. The area of the building is assumed to be 750 sq. ft., external walls are 25 ft. by 30 ft. by 8 ft. high, giving a wall surface area of 843 sq. ft. (This makes no allowance for door and window openings, but the amount over-estimated may be very roughly equated with the material wasted in mixing and placing.) If the walls are to be built 14 in. thick, the volume of the external walls will amount to 984 cu. ft. of compacted material, which will be equal to about 1633 cu. ft. of loose material. (Quoting Mr. Macdonald, *Terracrete*: "A cu. ft. of rammed wall will occupy about 1·66 cu. ft. as loose material, and will thus require ·01 × 1·66, or ·0166 cu. ft. of cement for a 1 per cent. mixture.") For a 5 per cent. mix (by volume), the amount of cement required would be about 82 cu. ft., or approximately 66 bags of cement containing 1 cwt. each. If 9 in. Stabilized Earth partitions were constructed, approximately another 24 cu. ft., or 18 bags, would need to be added to this total.

BITUMEN STABILIZED EARTH

THE term "Stabilized Earth" as applied to Adobe blocks is understood to mean earth stabilized by the admixture of a bituminous emulsion, or of tar. The process is called "Bitudobe" in America, where early in the history of "stabilization" its possibilities for building construction were appreciated, and numbers of bitudobe block houses were built in the south-western States. Only a small proportion of either bituminous emulsion or coal tar need be mixed with the earth in order to make it more resistant to water penetration.

It is believed that the action of the bitumen, like that of cement, is to block the capillary passages between the particles, and in this way to make the earth resistant to water penetration. In contrast to the action of cement, however, it appears that bitumen does not materially harden and increase the strength of the mix above the maximum strength of the soil itself. Nevertheless, when the right proportion of bitumen is used, there is no weakening effect, and by the added resistance to water penetration the strength of the soil is maintained at its maximum.

MATERIALS

Coal tar mixes readily with earth, but not all bituminous emulsions are as easy to use, and the intending builder would do well to consult a manufacturer as to the suitability of his products for use with soil. Bituminous emulsions require to be handled with care because the suspension of the particles in the liquid can be disturbed very easily. If this occurs, the particles will settle out, making the emulsion unworkable. The breaking-up can be caused by a change in the proportion of bitumen to water, such as can easily occur by evaporation, by freezing, or by the stirring of the emulsion with anything that will absorb water. A slow-breaking emulsion is recommended as being the best for use with earth, as it allows more time for mixing and moulding. Earths containing a high proportion of alkaline salts should not be stabilized with bituminous emulsions.

PROPORTIONS

The clay in the soil is the chief constituent which causes instability, and the amount of stabilizer used should be made proportional to the fine material contained in the soil. Mr. Edwin L. Hansen, in the *University of Illinois Bulletin No. 17*,[1] quotes a figure of 20 per cent. of the earth passing a 200-mesh sieve as being sufficient bitumen to stabilize most soils. The amount of fine material in the soil may be found by washing a weighed specimen of the soil over a 200-mesh sieve, drying it, and then weighing the remaining coarse material. From the difference between the weight of the coarse material and the original dry weight of the sample, the percentage of fine material can be found. The average amount of stabilizer used with a good soil is about 5 per cent. by weight of the dry earth—that is about 0·6 gallon per cu. ft. of dry earth.

MIXING

In order to obtain satisfactory results with three types of stabilizer, the bitumen must be very thoroughly mixed with the soil.

[1] Edwin L. Hansen, "The Suitability of Stabilized Soil for Building Construction," *University of Illinois Bulletin No. 17*, December 1941.

If done by hand the mixing will be found to be rather slow and laborious. Of mechanical mixers the pug-mill, or any type which has a slicing action, is the best, for the stabilizer tends to stick to the drum of the usual type of concrete mixer, and fails to mix satisfactorily with the earth.

Bituminous materials mix most readily with wet earth, and the following method of mixing has been found satisfactory: Of a given batch of earth, a quarter of the material is separated and mixed with the total amount of stabilizer and of water required for the whole batch. When these have been thoroughly mixed in, the remainder of the earth is added in dry. If the local soil is a heavy clay to which sand needs to be added to produce a suitable mix, it has been recommended that the total amount of bituminous stabilizer and of water required for a particular batch should first be mixed with the clay, and that the sand should then be added. In this way less mixing is required than if the sand were added before the bitumen, but the mixing should always continue until the material is of a uniform colour.

Moisture Content

Sixteen to 20 per cent. by weight is a suitable moisture content for mixing "Bitudobe" and moulding it into blocks, that is, the same moisture content as for plain Adobe blocks. More water makes the mixing easier, but has the disadvantage that voids are left in the blocks when the excess moisture has dried out, and the compressive strength is, therefore, not so high.

Curing

The blocks are cured in the same way as plain Adobe blocks; the longer they are cured the better.

Laying

They may be laid with a mortar made up from a thin mixture of the Stabilized Earth with the large aggregate removed, with a cement mortar containing bitumen stabilizer, or with a cement-lime mortar; mortar containing cement has the advantage of setting faster than a mortar made with earth only. The blocks should, of course, be bonded, as in other types of block construction.

COB, PISÉ, AND STABILIZED EARTH

WALL THICKNESS

The thickness of wall commonly used when building with Bitudobe blocks is 12 to 14 in. for one storey, and 12 to 20 in. for a two-storey structure. Chimney stacks have been built successfully with the blocks, but when this is done care must be taken to avoid exposing them to the fire or to any great heat.

RENDERING

The exterior may be rendered as in Pisé work. In addition, a treatment involving an aluminized asphalt paint has been recommended for undercoating and finishing, and is said to give good results. Internally, walls may be plastered, but other finishes, such as slurries which do not conceal the texture of the block construction, give an opportunity for imaginative interior treatment. (See Chapter VI for a discussion on protective coverings.)

Bitumen Stabilized Earth has not been used in this country as far as the authors are aware, but there seems no reason to suppose that buildings of this material would be any less successful than those which have already proved their worth in America.

RESIN AND OTHER STABILIZERS

As has already been explained, resins and oil stabilizers differ from Portland cement and bitumen emulsions in their mode of action. Furthermore, whereas some 5 per cent. or more of the former types of stabilizer are required, only a very small quantity, often less than 1 per cent., is required to waterproof the soil.

The action of the resins and waxy oils is thought to be roughly as follows:

The resins are substances having asymmetrical molecules, which are water attractive at one end and water repellent at the other, so that when powdered resin is brought into contact with water it forms a thin film over the surface in accordance with its polar structure, just as paraffin forms a thin skin when poured on to water. Each particle of soil is known to be surrounded by a film of water; when, therefore, resin is added, it forms a thin skin on the surface of each of these films in the earth mix, so hindering the intake of moisture and protecting the earth from rainfall and thus from instability.

The action of the waxy oils is similar, but the results are

Pl. 26.—Spreading the cement on the earth prior to mixing it with the earth with a disc-harrow.

Pl. 27.—The surface of a trial soil-cement wall being trowelled to a smooth finish.

Pl. 28.—Devon Country House, built of Devon Cob.

achieved by a slightly different means; here the wax is thought to form a layer of exceedingly small wax crystals on the surface of an oil film which in turn spreads over the surface of the water surrounding the soil particles. The film of wax crystals displays sufficient rigidity to prevent displacement of the film of oil by capillary force.

So far only a few stabilizers of these kinds have been marketed, mostly in America, and these have been restricted to essential work. However, those that already exist may later become available for general use and other synthetic substances may be developed, having essentially the same action. Those stabilizers already marketed are "Vinsol Resin" and "Stabinol," both produced by the Hercules Powder Co., of the United States, and "Shell Stabilizing Oil," produced by the Shell Group and first developed in Amsterdam. In England Soil Stabilization has attracted the attention of a number of firms, amongst them Imperial Chemical Industries Ltd.

Roads and runways of resin and oil Stabilized Earth have been constructed simply by disc-harrowing the soil, watering, adding the stabilizer, mixing in with the disc harrow, and finally compacting; this, of course, is possible only where the soil is suitably graded. Under the same conditions, earth for walling would be prepared in the same way, though, of course, it could equally well be mixed in heaps if this were more convenient or if the soil required additional material to obtain a suitable grading.

The resins and oils have one notable advantage over Portland cement, namely, that they are not affected by the weather, and earth stabilized with them is not harmed if it gets wet before being compacted. Of course, the moisture content of the soil should be gauged for ease of ramming before adding the resin. The amount of moisture should be about the same as in ordinary Pisé work.

As far as is known no buildings have ever been erected with this type of Stabilized Earth; but the method obviously has possibilities.

CHAPTER FIVE

Cob and Chalk Mud

I F ever the counties of England recover their bygone loyalty to their own materials and their old traditions, then Cob-building will return to Devon and the West. Cheap bricks, cheap transport, and the ignoble rage for fashions from the town went far to oust provincial Cob from the affections of those whom, with their forbears, it had housed so well for several centuries. For the soil of Devonshire, and of many parts of Wessex and of Wales, is excellent for building in Cob, known variously as "clom" or "clob," according to the locality. Innocent says,[1] " 'Cob' is a West of England word, and its literary history is short. The older form seems to be 'clob,' for a seventeenth-century Devon inquest speaks of 'a mudde or clobbe wall.' Whether it is derived from 'clob' or from 'korb,' the evidence in either case is in favour of a wattlework origin for the mud wall, as 'korb' means basket-work, and 'clob' and 'cleam' are forms of widespread Teutonic words for smearing on and plastering."

Cob walls are built of a mixture of clay, straw, and water. The mixed stuff is pitched on course upon course, without the use of shuttering. The faces of the walls are pared down as the work proceeds, and when they have dried out they are given some protective covering on the outside, and are plastered on the inside.

A mixture of crushed chalk and straw has also been used traditionally, with a technique similar to that of Cob walling. The Chalk Mud walling of Wiltshire, for instance, consists of well-trodden chalk lumps and straw, pitched upon the wall, and in Hampshire a Cob-type wall is made of three parts chalk to one of clay. It is paradoxical that the walls of chalk are known as "mud walls" and the labourers as "mud-wallers"; the traditional terminology is somewhat confusing.

The soil itself suggested the construction, and the men of Wessex were quick to take the hint and to act on it. The yeomen and small-holders of earlier days were commonly builders too,

---

[1] C. F. Innocent, *The Development of English Building Construction.* Cambridge University Press.

and often built their own homes in their own way, but guided by local tradition. Thus the old Devonian countryman in need of a house would set to and build it himself—of stone if that were handy and easily worked, of cob if it were not.

No doubt the doors and windows would be made and fitted by the village wheelwright; but the cottager himself would thatch or slate the roof as naturally and successfully as he built.

The skill and care with which these versatile amateurs built their houses was not always of the highest, and careless construction, like other sins, is visited on the children, the worse the sooner. Thus it is that there are today plenty of old Cob cottages that are both damp and insecure, but to condemn Cob building in general because certain old builders were careless, ignorant, or incompetent is to condemn all materials from wattle and daub to reinforced concrete in the same breath.

Cob, being a humble, amenable, and thoroughly accommodating material, has reaped the inevitable reward of good nature in being "put upon" and in being asked to stand what is quite beyond its powers of endurance, and yet Devon Cob houses of Elizabethan date are not uncommon.

It is very reasonable in its demands, but two things it does require—dry foundations and a good protecting roof. To quote an old Devonshire saw on Cob, "Giv un a gude hat and a gude pair o' butes an' er'll ast for ever."

In many instances the Devonshire leaseholder, usually only a "life-lease" holder, built badly and on indifferent foundations. He neglected to repair his thatch, with the consequence that ruin followed sooner or later. He did not always give the wall a proper protective covering, so that it often happened that by the time the lease expired the unfortunate landowner found that the cottage fell in—in the literal as well as in the legal sense. The lower portions of the walls were honeycombed with rat holes, the walls bulged out or were fissured by subsidence, and the dwelling presented that appearance of squalor and meanness that has led so many to decry the mud buildings of Devon as relics of a bygone barbarism. But if adequate care is bestowed on the construction, there is no reason why Cob cottages should not prove at one and the same time comfortable to the inmates and pleasant to the eye, and, moreover, endure for many generations.

Cob was not by any means exclusively the poor man's material, and several old houses of Cob still survive which are of some consideration. Amongst them is Hayes Barton, the birthplace of Sir Walter Raleigh. Writing of Raleigh and his home, Mr. Charles Bernard says:

"He had great affection for his boyhood's home—the old Manor House at Hayes Barton where he was born—and did his best to secure it from its then owner. 'I will,' he wrote, 'most willingly give you whatsoever in your conscience you shall deme it worth . . . for ye naturall disposition I have to that place being born in that house, I had rather see myself there than anywhere else,' but alas! it was not to be, and the snug and friendly tudor homestead passed into other hands. The house at Hayes Barton was probably not newly built when Raleigh's parents settled there, and it says much for its character that the house is as good today as ever it was; though for all that, it has, to use Mr. Eden Phillpotts' words, 'been patched and tinkered through the centuries,' it 'still endures, complete and sturdy, in harmony of old design, with unspoilt dignity from a far past.'"

In his *Cottage Building* C. B. Allen says:

"The cob walls of Devonshire have been known to last above a century without requiring the slightest repair, and the Reverend W. Ellicombe, who has himself built several houses of two storeys with Cob walls, says that he was born in a Cob-wall parsonage, built in the reign of Elizabeth, or somewhat earlier, and that it had to be taken down to be rebuilt only in the year 1831."

An old authority writes of Cob: "Walls of mud or compressed earth are still more economical than those of timber, and if they were raised on brick or stone foundations, the height of a foot or 18 in. above the ground, or above the highest point at which dung or moist straw was ever likely to be placed against them, their durability would be equal to that of marble if properly constructed and kept perfectly dry. The Cob walls of Devonshire, which are formed of clay and straw trodden together by oxen, have been known to last above a century without requiring the slightest repair; and we think that there are many farmers, especially in America and Australia, who, if they knew how easily walls of this description could be built, would often avail themselves of them for various agricultural purposes.

"The solidity of Cob walls depends much upon their not being hurried in the process of making them, for if hurried, the walls will surely be crippled, that is, they will sway or swerve from the perpendicular. It is usual to pare down the sides of each successive rise before another is added to it. The instrument used for this purpose is like a baker's peel (a kind of wooden shovel for taking the bread out of the oven), but the Cob parer is made of iron. The lintols of the doors and windows and of the cupboards and other recesses are put in as the work advances (allowance being made for their settling), bedding them on cross pieces, and the walls being carried up solid. The respective openings are cut out after the work is well settled." (It is interesting to note that this method of cutting out openings after the walls have settled is common practice also in those parts of Northern China where earth-walled houses are built.) "In Devonshire the builders of Cob-wall houses like to begin their work when the birds begin to build their nests, in order that there may be time to cover in the shell of the building before winter. The outer walls are plastered the following spring. Should the work be overtaken by winter before the roof is on, it is usual to put a temporary covering of thatch upon the walls, to protect them from frost."

The village and estate of Great Fulford can show as many good examples of old Cob work as any place in Devon. After the 1914–18 war Mr. Fulford of Great Fulford, whose name may be linked with Cob as one who has done much to promote its revival, wrote as follows:

"It is not possible to give a close estimate of what would now be the comparative cost of a building in cob, stone, or brick, as this must depend upon the exact locality of the site. It may, however, be of assistance if I quote particulars of the relative cost of cob and stone building in Devon in the year 1808 when cob was in common use. The stonework referred to was rough rubble, and not with square or dressed blocks. It must be borne in mind that up to that date practically all material, stone, lime, etc., was carried on horses' backs. Wheeled carts, which began to creep in about the beginning of 1800, were not in general use until twenty or thirty years later. As a boy I knew a farmer who remembered the first wheeled cart coming to Dunsford. In 1838 the Rector of Bridford (the 'Christowel' of Blackmore's novel) recorded the fact that in 1818 there was only one cart in the parish and it was scarcely used twice a year. In 1808 the price of building varied according to the district. In the northern part of the county the common price of stonework, including the value of three quarts of cider or beer daily, was from 22*d.* to 24*d.* the perch (16 ft.), 22 in. in width and 1 ft. in height. Including all expenses of quarrying and carriage of materials, stonework worked out at from 5*s.* to 6*s.* per perch running measure, and cob estimated in like manner at about 3*s.* 6*d.* Masons, when not employed by the piece, received 2*s.* per day, and allowance of beer or cider. In the Dunstone district (the clay shales from which make the best cob) masonwork was 18*d.* per rope of 20 ft. in length, 18 in. thick, and 1 ft. high, stone and all materials found and placed on the spot; cob work of the same measure was 14*d.* In the South Hams district masonwork cost 2*s.* 6*d.*, and cob 2*s.* per perch of 18 ft. in length, 2 ft. thick, and 1 ft. high.

"In those parts of the red land where Dunstone shillot or clay shale is not available, the red clay was mixed with small stones of gravel, and frequently the cob was laid and trodden down between side boards as used in building concrete walls. Three cartloads of clay built a perch and a half of wall 20 in. wide and 1 ft. deep. Eight bundles of barley straw, equal to one pack-horse load, were mixed and tempered with nine cartloads of clay.

"Thatching in 1808 cost 8*s.* per square of 10 ft.; 100 sheaves of wheat-straw reed, weighing 25 lb. each, were sufficient for one square. Thatching, however, is not, as many suppose, indispensable as a roofing for cob buildings; slate found in many parts of Devon was frequently used, and of late years Welsh and Delabole slates, tiles, and unfortunately, from the picturesque point of view, corrugated iron, have to a large extent supplanted thatch.

"Vancouver, in his report on the Survey of Devon for the Board of Agriculture in 1808, gave the following recipe, which he described as a preserving and highly ornamental wash for roughcast that was then getting into common use: 'Four parts of pounded lime, three of sand, two of pounded wood ashes, and one of scoria of iron, mixed well together and made sufficiently fluid to be applied with a brush. When dry it gives the appearance of new Portland stone, and affords an excellent protection against the penetrating force of the south-westerly storms.'

"Cob-making was, like many other local trades, carried on in some families from generation to generation and developed by them into an art, but apart from these specialists, practically every village mason and his labourers built as much with cob as they did with stone. There are men still left in various parts of the county who have made cob, and it would, in my opinion, be of advantage if demonstrations could be given by them to discharged sailors and soldiers who are anxious to take up work on the land.

"The depletion of our home-grown timber supply and the prohibitive cost of practically all building material has in effect brought about the conditions that led our forefathers to utilize suitable material that lay nearest to hand, and unless some endeavour is made to follow their methods and profit by their example, it will be impossible to provide sufficient buildings for the necessary equipment of the allotments and small holdings, let alone housing accommodation for the workers on the land."

Mr. S. Baring-Gould, in his *Book of the West*, writing on the subject, says: ". . . I have known labourers bitterly bewail their fate in being transferred from an old fifteenth or sixteenth-century cob cottage into a newly-built stone edifice of the most approved style, as they said it was like going out of warm life into a cold grave." And of garden walls he says: "Cob walls for garden fruit are incomparable. They retain the warmth of the sun and give it out through the night." [This is an interesting testimony to the high heat capacity of earth walls.] ". . . when protected on top by slates, tiles, or thatch, will last for centuries."

It is not intended to argue that the Cob cottage could be advantageously built in every county, but only that where it has been used and liked for centuries, as in Devon, a wise building policy would encourage its continuance. The materials are at hand, and there are many who would readily welcome its revival.

This is true, also, of the Chalk-mud houses of Wiltshire. The village of Winterslow has given its name to a particular method of chalk-mud walling practised in the district, which is similar to Cob, and from *Country Life*, April 6th, 1901, we read:

"The white chalk cottages of the scattered straggling village are found in every sort of position. They must not be confounded with the cottages of rock chalk at Medmenham. You might almost call them mud cottages.

"The house is generally both planned and constructed by the owner.

". . . The soil is only a few inches deep, soft chalk lies close to the surface and can be dug out with a spade. This is a very suitable material in the district and costs nothing but the labour of digging. . . .

"On the downs there is a constant lack of water; that which falls in the shape of rain is therefore very precious, and in some cases is indeed the only kind available. But a large tank or artificial well is needed to contain it, and the pit from which the chalk is dug can be made to serve the purpose. . . . One was made watertight by means of a lining of concrete, and held enough water to keep the family going through all the dry season.

"In another house . . . the chalk-pit had been utilized to form a large and convenient cellar. . . .

"Most of them [the cottages] . . . are on two floors, with parlour, kitchen, back kitchen and so forth on one, and the bed-rooms on the other. In the preparation of the chalk, the method followed is that of treading it into a kind of rubble, and adding a proportion of straw and a small quantity of lime.

"There is a local builder who will run up the shell of a house for a matter of £100, more or less, according to its size. . . . Most of the cottages are literally hand-made. A skilful architect who visited the Winterslow cottages felt sure that boards must be used to keep the walls straight, but he was wrong. The chalk is shovelled up and the walls are kept straight without line or plummet. No expensive scaffolding or machinery is employed. Yet the walls come out beautifully in the end, the colour being an exquisite soft white. They are about 18 in. thick, and the slow-ness of their construction has one good effect, it gives them time to dry. No point is of more importance than this. It is advisable not to put on any rough-cast, plaster, or paper for at least twelve months, as doing so will prevent the moisture from exuding. One or two of the little cottages were slightly damp, but the majority were as dry as tinder. The thickness of the walls helps to render the cottage more comfortable, to make it cool in summer and dry in winter.

"One word should be added in regard to soft chalk as a building material. Where it can be obtained in the garden at a few inches depth, and especially where the cottager is his own architect and builder, it can be most heartily recommended, but there are obvious objections to its transportation to districts where it is foreign.

"The village itself is a very homely and irregular one without a

Pl. 29.—Elizabethan Cob House, Lewishill. Walls from 3 ft. to 4 ft. thick. A wing was added in 1618. This farm has been occupied by the family of the present holder for between three hundred and four hundred years.

Pl. 30.—A Devon Cob Farmhouse, probably between two hundred and three hundred years old.

Pl. 31.—Another view of the Cob House built by Ernest Gimson near Budleigh Salterton, Devon.

Pl. 32.—Three Chalk Cottages at Hursley Park.

Pl. 33.—The same Cottages in 1945, twenty-five years after building.

single dwelling of any pretence. The country lying adjacent to Salisbury Plain consists of broken, sparsely peopled downland, and very ornate or finished cottages would be out of keeping, but they would not look so well copied in a very rich, heavily timbered country."

THE TECHNIQUE OF BUILDING IN COB

COMPOSITION

Local custom as to the composition and preparation of the mixture will generally be found to have adjusted itself to the peculiarities of the soil.

The following is an analysis of a sample of typical old Cob walling:

							Per cent.
Stones (residue on 7 by 7-mesh sieve)		.			.		24·4
Sand, coarse (residue on 50 by 50-mesh sieve)		.					19·7
Fine sand (through 50 by 50-mesh sieve)		.				.	32·5
Clay	.	.	.	.	.	.	20·6
Straw	.	.	.	.	.	.	1·25
Water, etc.	.	.	.	.	.	.	1·55

The analysis shows that proper grading, as for good Pisé earth, is necessary for Cob walls. The moisture content, however, needs to be higher and the material more plastic than for Pisé, because of the walling technique peculiar to Cob. With Cob, the walls consolidate by virtue of their own weight and by the action of capillary forces as the water in them dries out. As with Pisé the clay encourages cohesion between the earth particles, and sand helps to produce a stable wall. When dried out, a Cob wall is very similar to a rammed earth wall, but in the process of drying out considerable shrinkage will have taken place, and it may be that a function of the straw admixture is to divide up the planes of contraction, and so localize any tendency to cracking.

Lime has sometimes been added to the clay or chalk mixture, and it will have been seen from the foregoing chapters that an addition of lime or preferably cement can be used in earth construction to produce a stronger wall.

As a general rule, when building in Chalk Mud, the finer the chalk the stronger and more durable is the walling.

Some of the old builders seem to have been somewhat catholic in their conceptions as to what constituted "chalk," and vague patches of earth, loose flints, and other stray substances not infrequently mar their work and sometimes seriously reduce its strength. What is aimed at is a conglomerate of small chalk knobs cemented together by a matrix of plastic chalk and straw, the whole forming as dense a mass as possible.

MIXING

The old method of mixing Cob by hand is as follows: A "bed" of clay-shale is formed close to the wall where it is to be used, sufficient to do one perch. A perch is superficial measurement described as 16½ ft. long, 1 ft. high, and the amount of material will vary according to the thickness of wall required. Four men usually work together. The big stones are picked out. The material is arranged in a circular heap about 5 or 6 ft. in diameter, and starting at the edge the men turn over the material with Cob picks, standing and treading on the material all the time. One man sprinkles on water and another sprinkles on barley straw, from a wisp held under his left arm. The heap is then turned over again in the other direction, treading continuing all the time. "Twice turning" is usually considered sufficient. Straw bands may be wrapped puttee-wise around the legs of the men to keep them clean, and these are removed at the end of the day.

More rarely the mixing is done in a rough trough, whilst a power-driven "pan-mill" has also been tried with success; though one would think that the use of such a machine might tend to diminish the binding strength of the straw submitted to its grinding.

CHALK MUD

The traditional method of mixing Chalk Mud differs but slightly from that of mixing of Cob.

The top soil is removed and put aside for subsequent garden use, and the chalk is then dug, little or no attempt being made to break up the larger lumps still further. Very often the chalk is dug in the autumn and the frost allowed to play on it during the winter. In this way the chalk is broken up by frost action and the natural process helps to achieve the grading necessary for a strong and stable wall. A platform of levelled ground is then prepared around the outside of the walls and the chalk loosely spread on to it. Straw is sprinkled on and the whole is then well trodden, usually by the labourers but sometimes by horses. The quality of the walls depends very largely on the preparation, that is, on getting the mud to the right consistency, and the old hands know by experience when it is ready.

BUILDING

A foundation wall of flints or brick or rubble stonework, known sometimes as an "underpin" course, is carried up to a height of about a foot above ground level, or sometimes to the height of the lower-floor window cills, and is finished to a plane surface to re-

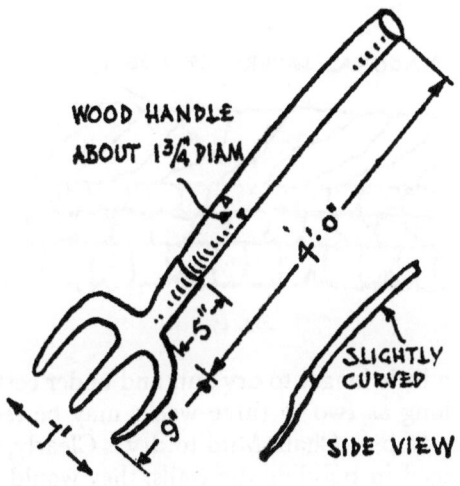

COB PICK

FROM EXAMPLE · AT GREAT FULFORD

WOOD HANDLE ABOUT 1¾ DIAM

4'·0"

5"

1'

9"

SLIGHTLY CURVED

SIDE VIEW

Fig. 16.

ceive the mixed stuff. A damp-proof course should always be included in this plinth wall.

A waller stands on the wall, armed with a three-pronged pick, not unlike a trident (Fig. 16). His assistants hand up a mass of the material on the end of a similar pick, and the waller receives it and lays it on the wall, treading it into position. Thorough treading is important, and the heels should be well used. The material is allowed to project each side an inch or so beyond the plinth wall to allow for the paring which follows, as the courses are completed.

The courses are usually from 18 in. to 2 ft. high, and the material is laid and trodden in diagonal layers as shown in the diagram (Fig. 17), thus securing proper bonding. The walls should be covered at night in case of rain, and new work, especially in Chalk Mud, must not be exposed to frost, or there will be danger of collapse.

Chalk Mud is sometimes sufficiently solid by the time the waller has returned to his starting-point, having completed a course round the building, for him to commence upon the next course.

COB COURSE OR SCAR

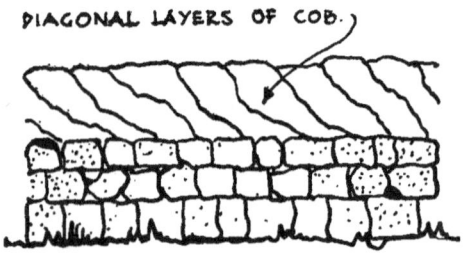

Fig. 17.

Cob may take a little longer to dry out, and under certain weather conditions as long as two to three weeks may be required for a course of either Cob or Chalk Mud to dry. Clearly, if five or six men were engaged in building the walls, they would not then be continuously employed. It would be of advantage, therefore, to have several buildings in hand at the same time, so as to be able to turn from one to the other while the courses were drying.

At the completion of a course, the corners are plumbed up from the plinth wall; a line is stretched through, and the face of the wall is pared down to a vertical surface with the "paring iron" (Fig. 18) by the man standing on the wall. Four men will do about four perches of a wall 2 ft. thick in a day, preparing and laying the material.

The material is rarely laid between timber shuttering as in Pisé work, as the retaining boards tend seriously to retard the drying out.

DRYING OUT

The foregoing paragraphs clearly stress the necessity for thorough drying out. The walls should be built from about March to September, but the internal fitting, plastering, etc., can be done in the winter. The external rendering must not be done

PARING IRON FOR COB WALLS
MEASURED FROM EXAMPLE AT GREAT FULFORD

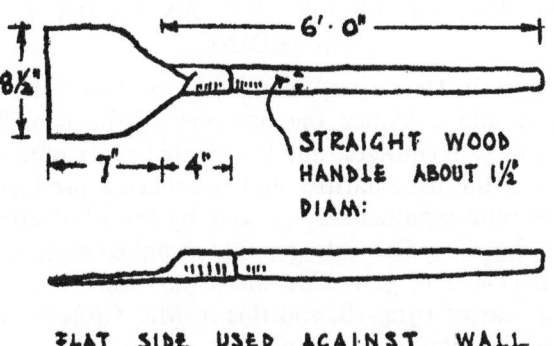

FLAT SIDE USED AGAINST WALL

Fig. 18.

for at least a year, perhaps two years, to allow the walls to become thoroughly dry.

As unprotected Cob and Chalk Mud are sensitive to frost and rain, they should be given a good external protective coating as soon as they are really dry, and should, in the meantime, be protected by a temporary covering of some kind such as straw matting. All Cob and Chalk Mud work should also be protected from frost and rain during the building operations.

No artificial methods of drying out are at present usual, beyond good fires inside during the winter; though, as under such conditions a Cob cottage is not usually considered fit to live in for several months after completion, some artificial means of drying might be worth considering.

Design

Many of the design features of Cob and Chalk Mud are common to all forms of earth building, and an account of them has been given in Chapter VII. The reader will also find information on protective coatings and renderings in Chapter VI.

Clay Lump

Clay lump is a "Cob" block which is but a form of Adobe, and as such is discussed in Chapter III, where the reader will find an account of its use in this country.

SOME EXAMPLES OF COB AND CHALK MUD BUILDING

A DAY's walk in Devon, or, failing that, a glance at the printed pictures, should convince the observer of the comeliness and durability of Cob cottages; and that their beauty does not derive merely from the irregularities and weathering produced by the passage of time is sufficiently proved by the illustration of Mr. Gimson's charming Cob cottage, photographed soon after he had finished it (Pls. 1 & 31). The work was done a year or two before the war of 1914–18, and this is Mr. Gimson's own description of the manner of its building.

Mr. Gimson's Cob House

"The Cob was made of the stiff sand found on the site; this was mixed with water and a great quantity of long wheat straw trodden into it. The walls were built 3 ft. thick, pared down to 2 ft. 6 in., and were placed on a plinth standing 18 in. above the ground floor, and built of cobble stones found among the sand. The walls were given a coat of plaster and a coat of rough-cast, which was gently trowelled over to smooth the surface slightly. I believe eight men were engaged on the Cobwork, some preparing the material, and others treading it on to the top of the walls. It took them about three months to reach the wall plate; the cost was 6s. a cubic yard, exclusive of the plastering. No centering was used. The joists rested on plates, and above them the walls were reduced to 2 ft. 2 in. in thickness to leave the ends of the joists free. The beams also rested on wide plates and the ends were built round with stone, leaving space for ventilation. Tile or slate lintels were used over all openings. The cost of the whole house was 6½d. a cubic foot. Building with Cob is soon learnt—of the eight men, only one of them had had any previous experience, and, I believe, he had not built with it for thirty years. This is the only house I have built of Cob."

What is most interesting in this narrative is the workmen's lack of experience, which seems to have been no hindrance. Anyone who proposes to revive the use of Cob may take courage from Mr. Gimson's evidence. The time spent in building the walls was reasonable and the cost low. It may be guessed that the post-war rise in cost will be no greater in proportion, if as great, when compared with brickwork. The natural charm of the wall surface is enhanced by the crown of thatched roof, modelled with a skill which few can bring so certainly to their task as Mr. Gimson (Pl. 1).

Just as there are many Cob buildings in Devon, so are there Chalk-mud buildings in such a typical chalk district as that lying about Andover in Hampshire, and those who may wish to see buildings in Chalk Mud, both old and new, would do well to visit some such place.

It should, however, be constantly borne in mind that most of the old cottages were somewhat unscientifically erected by their original jack-of-all-trades occupiers, that damp-proof courses and Portland cement were unknown, and that the advantages of proper ventilation and the causes of dry-rot were discoveries yet to be made. Also, a large number of these cottages have been sadly neglected either recently or in the past, and they bear the disfiguring marks of their ill-treatment upon them.

But a chalk cottage that is well found in the beginning, and is reasonably well cared for subsequently, has nothing to fear from comparison with cottages built in the most approved manner of the more fashionable materials.

THREE CHALK-MUD COTTAGES AT HURSLEY

Mr. James Thorold gives the following particulars of a block of three chalk cottages built for Sir George Cooper on his estate at

Hursley, near Winchester. Pl. 33 shows the cottages as they were in 1945, some twenty-five years after building.

"The chalk walling was done by Messrs. A. Annett & Son, of Winterslow, near Salisbury, where this method of building has been kept alive from olden days. It consists of working up the soft upper strata of the chalk by putting a bed of it 4 ft. 6 in. thick on the ground, watering and treading it to a sticky consistency with the feet, working in shortish straw at the same time. When thoroughly mixed by the builder's mate, he lifts up a forkful to the builder working on the wall immediately above him, the latter catches the chalk, dumps it down on the top of the wall, building an 18-in. course all round. As soon as the weather has dried this sufficiently, he goes round with a sharp spade squaring up both sides of the wall. As this work is greatly dependent on the weather it is well if the men have other work to fall back on, and that building operations should be commenced in the spring or early summer. The wall is built 18 in. thick to the first floor joists and 14 in. above. Chalk in itself being very absorbent of moisture, the usual plan is to render the outside of the wall with a lime mortar, which, however, requires renewal every few years. To obviate this we fixed with long staples 1¼-in.-mesh wire-netting over the outside surface of the wall to give a reinforcement for a rendering of hair mortar and cement gauged in proportion of 1 to 2 respectively, and left rough from the trowel. This rendering was done at a cost of 3s. 3½d. per square yard, which is a substantial addition to the cost of the walling, but so far there is no sign of a crack or hollow place behind it, and the cottages have kept dry. The walls were finished off with a limewash containing Russian tallow and copperas.

"As regards the cost of this block of three cottages, the result is obscured by the fact that tall chimney stacks with ornamental bricks and appropriate foundations were built and reinforced leaded lights were used in the windows to keep the building in character with the other cottages on the estate, but at the time we estimated that the chalk walling saved a sum of £54 as against the amount we should have had to spend in carrying out the building with bricks made on the estate, and this had to include lodging money and profit, the builders being independent men. The ornamental chimney stacks were put in for the sake of appearance, flues built up in the chalk being entirely satisfactory and

Pl. 34.—Chalk Cottages at Hugh's Settlement, Quarley, Hampshire. Successful experiments have been made at Quarley in chalk building dating back to 1923. The work was interrupted by the second World War, but further experiments are now in progress, under the guidance of Mr. B. H. Nixon and of Miss Jessica Albery, A.R.I.B.A. With the experience already gained, the results should be especially valuable.

Pl. 35.—A further example of Chalk Building at Hugh's Settlement. The illustration shows a house by Miss Albery under construction. The walling is of pre-cast chalk blocks covered externally with a lime rendering.

ROBERT O. MENNELL'S CHALK BUNGALOW

Pl. 36.—Crushing Chalk in a cattle-cake cutter. The tarpaulin covering is to enable the men to work in wet weather.

Pl. 37.—Boarded roof covered with bitumen sheeting ready for slating, and walls before being rendered.

Pl. 38.—Showing foundations, damp course, chalk wall, and flint facing.

Pl. 39.—Plinth of split flints and tile creasing.

Pl. 40.—Smoothing the exterior as shuttering is removed.

fireproof. The foundations are either flint or brick with a slate damp-course.

"I consider that for a chalk country this method of building has many advantages.

"(1) It saves cartage.

"(2) It can be carried out by a skilled labourer who can be otherwise employed during unsuitable weather.

"(3) No fuel is required as in burning bricks.

"(4) If a suitable rendering is employed to keep it weatherproof, and a good damp-proof course on the foundations, the cottages are nice and dry and keep an equable temperature, chalk being a good non-conductor."

MR. MENNELL'S CHALK BUNGALOW

In *Country Life* (September 10th, 1943), Mr. Robert O. Mennell described in some detail how he built a bungalow in 1920 in chalk, which was designed by Mr. Geoffrey Morland. The method used was not that used traditionally for Chalk Mud since the walling material was, in fact, tamped down between shutters. This is Mr. Mennell's own description of the cottage and its building:

"Situated at Kenley in Surrey at an altitude of 500 ft., my bungalow stands on the brow of a hill and catches the full force of rain, wind and sun from south-east round to due west. To build with unconventional and, so far as our day and generation are concerned, untried material therefore required some degree of daring. Measuring externally 45 ft. 6 in. by 28 ft. 3 in., the house covers an area of 1,300 sq. ft. and cubes out at about 20,000 ft. The accommodation consists of a living-room, 21 ft. 6 in. by 13 ft. with a large bay window, four other rooms, a scullery-kitchen, a spacious bathroom, and a large attic lit by a dormer window.

"My own knowledge of building construction was sketchy in the extreme, but my joiner friend and his 19-year-old son, working with a stalwart navvy and an unemployed milk roundsman (we thought he would probably know as much as anyone about mixing chalk and water), made a fine team, and there was no necessity for a building contractor. Between them these four men not only cleared the site, dug the foundations and constructed the walls and roof (except for the slating); but made and erected the concrete blocks forming the interior partitions. Mr. Poynter himself did all the joinery work—doors and window frames, wardrobes, bookshelves and cupboards built into the house. The usual tradesmen, a bricklayer for the chimney stack, a plasterer and a plumber, were called in as and when required for their specialist jobs, and a firm of slate merchants laid the slates.

"After the site had been plotted and cleared, digging the chalk was commenced at a distance of 15 ft. from the southern end.

"When the walls were completed, the pit from which the chalk had been quarried measured 20 ft. square by 12 ft. deep. This pit, filled up with the turf and soil cleared from the site, eventually became an excellent vegetable plot.

"Some of the chalk was almost as hard as limestone, so that the job of breaking it down to 2-in. size or less proved more difficult than had been anticipated. Eventually a cattle-cake cutter solved the problem, but it meant hand milling, a somewhat tedious and relatively costly method. Actually, the chalk would have been more easily crushed if the disintegrating top layer had been taken, or if it had been quarried in the autumn previously and exposed to a winter frost. After quarrying and milling, the method of preparation was to take a cubic yard at a time, throw in a double-handful of short wheat straw as a binding element and turn the whole over with a shovel three times, playing a garden hose on it the while. If too dry, chalk cob will not bind unless rammed; if too wet, it will sag and bulge, and cracks may develop while the walls are drying out."

Foundations were constructed 3 ft. wide and 1 ft. high and were brought a few inches above ground level. On these a bitumen damp-proof course was laid, before commencing the walls.

Since wet chalk, broken small, binds readily and stands up well, neither elaborate shuttering nor expensive rammers were found necessary.

"In my case a few scaffold boards were found to be quite adequate and were moved along for each day's work. To hold them in position, 4 in. by 2 in. rafters, afterwards used for the roof, were fixed as upright struts and braced from the outside. The chalk was shovelled in between the boards and well punned to avoid air-pockets. An 18-in. lift was taken right round the building, so that by the time the men were round to the starting-point the first lift had had time to set hard.

"As each day's work ended a diagonal ramp was left to help

COB AND CHALK MUD

splicing on new work the following day. At night the walls were protected against rain by the placing over them of bitumen sheeting, which was eventually laid under the wall plate as an upper damp-course. For the first foot above ground the chalk was faced with a plinth of split flints to a depth of 4 in., this being surmounted by a tile creasing. This flint facing and the tile creasing not only stop erosion due to rain splashes or surface water, but very definitely improve the appearance of the building. They also counter penetration by rats or other vermin. Near one corner where two door openings and a window were in close proximity, it was deemed advisable to use brick jambs for the doors and reinforced concrete lintols with a 10-in. bearing on each side.

"Chalk absorbs moisture, and for this reason it is essential to provide a good 'skin' to prevent trouble from frost. On the other hand the sun dries moisture out of the chalk when laid wet, and for this reason the exterior rendering must not be impervious. Cement or paint cannot therefore be used, for the wet would collect inside the rendering and tend to 'blow' it.

"At Kenley we made a lime mortar, using sand found on the site, linseed oil, and horsehair; the chalk, brushed free of loose particles, forming an excellent natural key. A lime and tallow wash was added later; the mixture, 1 lb. of tallow in a bucket containing one-third quicklime and two-thirds water, was applied hot immediately after mixing and straining. The pure white finish of this wash is appropriate to a chalk house; it has proved most effective in protecting the walls from weather and has shown no tendency to flake off.

"The fixing of creosoted wooden window frames presented no difficulty. Though the walls were built up wet round them, there has been no sign of decay or dry rot anywhere. External cills, consisting of a double tile creasing, and internal cills of 1½ in. red half-glazed facing bricks covering the thickness of the walls, have proved both effective and attractive.

"In view of the exposed position I determined to make a thoroughly sound job of the roof. Three-quarter-inch boarding was laid on rafters set 12 in. apart; then a covering of bitumen sheeting and, over all, best quality green Norwegian slating 18 in. by 10 in. laid to a 3-in. lap and finished with half-round ridge and hip tiles. This work was done by a firm of slating

99

contractors at a price of £140. Admittedly mine is a very heavy roof, but an 18-in. chalk wall can carry any ordinary load even when taken to an upper storey. As chalk walls will not stand lateral thrust, the roof was well tied, so that in effect it rests on the building like a lid. Acting upon the advice of an old Devon saying applied to chalk houses, 'Giv un a gude hat and a gude pair o' butes and er'll last for ever,' we provided not only stout bitumen damp-courses both above and below the chalk, but also a good overhang of 18-in. eaves.

"For two open fireplaces a central chimney stack was erected not contacting the chalk at any point, lest the heat might adversely affect the cohesion of the Cob. For gas-fire flues in the other rooms 4-in. diameter iron piping was set in the chalk walls during construction.

"The saving effected by using materials found on the site and thus involving neither haulage costs nor manufacturers' or merchants' profits needs no emphasising.

"My one-acre field produced not only the chalk for the walls but also gravel, sand, and flints, all within a distance of 36 paces of the house itself. The gravel we used coarse for the foundations and sifted fine for the lintols and for the blocks used as interior partitions. Sharp clean sand found near the gravel pockets was used for both the interior plaster and the exterior rendering, while flints, which were plentiful in the chalk, made an ideal anti-erosion plinth. On any typical English countryside where chalk is the subsoil a similar combination of raw materials may be expected as more or less the normal thing. We calculated that we saved haulage on at least 250 tons of material, to say nothing of the trade profit thereon. Despite the fact that we hand-milled our chalk and were experimenting as we went along, the 18-in. chalk walls worked out at a cost of only 4s. 6d. a yard super as compared with 12s. 3d., the then current cost of 14-in. brick. This low cost of chalk construction is due to the fact that there is only digging, milling, and laying involved at an unskilled labourer's wage rate. If groups of chalk houses were built at a time it would pay to install power-driven plant for lifting, crushing, and mixing, and no doubt costs would thus be considerably reduced. Our all-in cost for the house worked out, as nearly as we could estimate, at 4d. per cubic foot less than brick houses, despite the fact that we put on the very substantial roof described, used only well-seasoned timber throughout, and made all doors and window frames by hand.

"In the experiment I have described neither the architect nor the man in charge nor any of the operatives had had any previous experience of building in this material, yet they were able to make a complete success of the job. In our experience no single crack

EXAMPLES OF WALL COPINGS

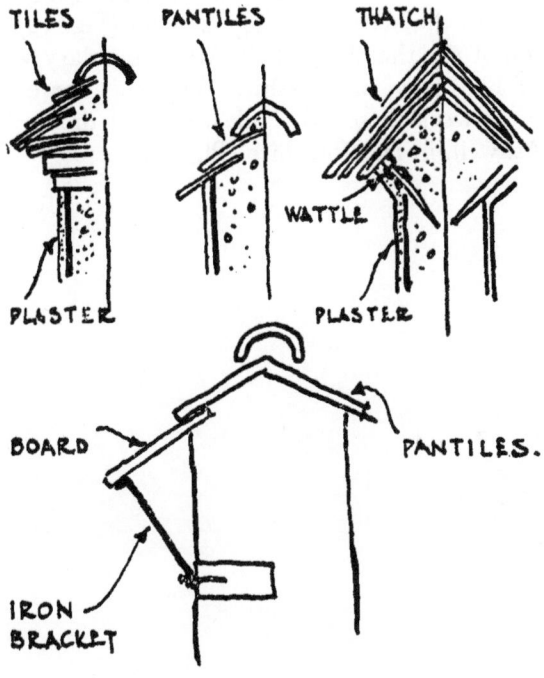

TILES PANTILES THATCH

WATTLE

PLASTER   PLASTER

BOARD   PANTILES.

IRON BRACKET

Fig. 19.

has developed and the walls have given no trouble whatever. By contrast my house alongside, constructed 10 years earlier of 14-in. brick rendered with 1-in. cement and sand, gave much trouble owing to the exposed situation and had eventually to be tiled on the sides open to south and south-west to keep out the weather.

"On the basis of a pound of experience being proverbially worth a ton of theory, I offer this detailed account of my own experiment as a contribution towards the joyous task of post-war reconstruction." (Pls. 36–40.)

GARDEN WALLS

Chalk, like Cob, has been used traditionally for the construction of garden and boundary walls. The "hat" with which they are provided is of the highest importance to the health and longevity of the walls, for they need as much protection as the walls of a house, and on both sides.

The appropriateness and the beauty of some traditional copings can be seen in the accompanying sketches (Fig. 19), and in the illustration of the thatched Chalk Mud boundary wall from the village of Blewbury, in Berkshire (Pl. 42). There is a striking similarity between these old chalk walls and those of chalk and clay at Ashwell, in Hertfordshire (Pl. 41).

CHAPTER SIX

Protective Coverings for Earth Walls

IN a tropical or semi-tropical climate the exterior of an earth wall becomes partially baked by the sun, and the hardened surface will often give sufficient protection to keep the wall in good condition without any other covering, or with only a mud plaster. For instance, in *A Manual on Earthwork*, edited by Colonel Maclagan, R.E., it is stated of earth walls in India that:

"Without plaster, good Pisé work is found successfully to withstand exposure to the weather, and after the lapse of many years to be so compact and hard as to be picked down with difficulty."

However, in a temperate climate, the providing of some form of external protection is important, for without it earth walls will, under extreme conditions, disintegrate and revert to mud. A protective covering is also specially necessary for walls of dwelling houses where rain penetration and dampness must be eliminated as completely as possible, not only to ensure that the walls are stable, but also to protect the health of the occupants.

Nevertheless, walls made with a properly graded earth, and especially with Stabilized Earth, will often remain in surprisingly good condition for many years when left unprotected, despite an initial weathering of a rather severe character which generally takes place in the first year or two. The walls of the little fruit-house built by Mr. St. Loe Strachey, and described and illustrated (Pls. 2–5) in his introduction to this book, are an example of Pisé which has endured without any protective covering, although it must be admitted that the generous eaves overhang and the relatively sheltered position in which the hut stands have undoubtedly contributed to their preservation.

A TRADITIONAL METHOD

A common traditional external protection was a rendering of lime roughcast. The following extracts by Dr. Abraham Rees from the *Cyclopædia or Universal Dictionary of Arts, Sciences and Literature*, from which much has already been quoted in the chapter on Pisé, gives a clear picture of one of the common ways in which this was done:

"To prepare the walls for plastering, indent them with the point of a hammer, or hatchet, without being afraid of spoiling the surface left by the mould; all those dents must be made as close as possible to each other, and cut in from the top to bottom, so that every hole may have a little rest in the inferior part, which will serve to retain and support the plaster.

"If you happen to lay the plaster over them before the dampness is entirely gone, you must expect that the sweat of the walls will cast off the plaster.

"The wall surface having been duly hammer-chipped, the work must be scoured with a stiff brush to remove all loose earth and dust, and finally to prepare it for roughcasting. Roughcast consists of a small quantity of mortar, diluted with water in a tub, to which a trowel of pure lime is added, so as to make it about the thickness of cream.

"One workman and his labourers are sufficient; the workman on the scaffold sprinkles with a brush the wall he has indented, swept, and prepared; after that he dips another brush, made of bits of reed, box, etc., into the tub which contains the roughcast, and with the brush throws the roughcast against the wall.

"Roughcast, which is attended with so little trouble and expense, is, notwithstanding, the best cover that can be made for Pisé walls, and for all other constructions; it contributes to preserve the buildings. It is the peculiar advantage of these buildings that all the materials they require are cheap, and all the workmanship simple and easy."

TRANSPARENT COVERINGS

Bare earth walls of Pisé, Cob, or Chalk Mud are often very beautiful; small stones in a Pisé wall give it the appearance of brown marble, and the beauty of the soft white colour of newly finished chalk walls has often claimed attention. Thus it has always seemed regrettable that the identity of these walls should be concealed behind an external rendering, and many experiments have been made to discover some transparent covering, which will give sufficient protection, but which will not hide the walling material.

Unfortunately these experiments have not been very successful.

One of the materials which has been tried is sodium silicate, commonly known as waterglass, but this usually fails a short time after it has been applied, and, in addition, is said to damage the walls by penetrating and then causing the earth to flake away. Various types of oil have also been tried, both to prevent the tendency of the earth to dust, and to act as waterproofers, but their effective life is short. There is as yet no evidence to show that the proprietary colourless waterproofers would be effective on earth walls.

LIMEWASHES

In the absence of a suitable transparent covering, perhaps the simplest and most appropriate treatment which does not completely conceal the identity of the wall is a limewash. It is easily applied with a brush, and with suitable pigments may be attractively coloured. There is, however, a certain prejudice against its use, because of the frequent renewal which is necessary: a new application is required every year or two; but if the building be of one storey only—and this is perhaps most suitable for earth construction—the whole wall surface is within easy reach without the use of ladders, and the task of renewing the wash should not be a difficult one. Frequent renewal has the advantage that the building retains a fresh and cheerful appearance, and with a limewash the cost is trifling.

The basis of most limewash recipes is the mixing of a quantity of tallow, which may be from 2 to 10 lb., into a bushel of quicklime to form an insoluble calcium soap. The tallow should be placed in the centre of the quicklime and the whole should be slaked together. If the quicklime is slow in slaking, it should be covered with sacking, and hot water should be used. The addition of pigments may necessitate an increase in tallow, but a useful mean to remember is 5 lb. tallow to a bushel of quicklime.

When tallow is not available calcium stearate in powder form may be substituted, or linseed oil may be added. Pigments are useful in giving opacity to the limewash, which would otherwise be transparent when wet with rain. A small quantity of carbon black making a pale grey wash is sufficient for this purpose. The pigments should, of course, be limefast, and should be added during slaking. If this is not possible the pigment should be mixed with alcohol and added to the strained whitewash. The covering power of these washes is about 350 to 400 sq. yds. per cwt.

There are several traditional formulæ consisting of lime (not whiting) thoroughly slaked and thinned to a cream to which various additions are made, such as salt, alum, powdered glue, casein (skimmed milk), etc. The effect of salt is probably to hold the moisture and facilitate the carbonation of the lime, while the addition of a small quantity of alum improves the working qualities and is thought to increase the hardness of the surface. Caseins and glues give greater binding properties to the mix.

Three typical recipes are given below; the choice of mix must depend upon the type of wall and the degree of exposure. Generally, the strongly bound mixes should be applied over the harder wall surfaces, the weaker mixes being suitable for porous and slightly rough grounds. Those containing glues should never be applied over tar or bituminous waterproofers.

The following recipes are taken from *White Paints and Painting* (Scott), and are reliable:

(1) *"Factory" Whitewash (interiors), for Walls, Ceilings, Posts, etc.:*

(*a*) 62 lb. (1 bushel) quicklime, slake with 15 gallons water. Keep barrel covered till steam ceases to arise. Stir occasionally to prevent scorching.

(*b*) 2½ lb. rye-flour, beat up in ½ gallon of cold water, then add 2 gallons boiling water.

(*c*) 2½ lb. of common rock-salt, dissolve in 2½ gallons of hot water.

Mix (*b*) and (*c*), then pour into (*a*), and stir until all is well mixed. This is the whitewash used in the large implement factories, and recommended by the insurance companies. The above formula gives a product of perfect brush consistency.

(2) *"Weatherproof" Whitewash (exteriors), for Buildings, Fences, etc.:*

(*a*) 62 lb. (1 bushel) quicklime, slake with 12 gallons of hot water.

(*b*) 2 lb. common table salt, 1 lb. sulphate of zinc, dissolved in a gallon of boiling water.

(*c*) 2 gallons skimmed milk.

Pour (*b*) into (*a*), then add the milk (*c*), and mix thoroughly.

(3) *"Light House" Whitewash:*

(*a*) 62 lb. (1 bushel) quicklime, slake with 12 gallons of hot water.

(*b*) 12 gallons rock-salt, dissolve in 6 gallons of boiling water.

(*c*) 6 lb. of Portland cement.

Pour (*b*) into (*a*), and then add (*c*).

Note.—Alum added to a lime whitewash prevents it rubbing off. An ounce to the gallon is sufficient.

Flour paste answers the same purpose, but needs zinc sulphate as a preservative.

PAINTS AND SLURRIES

In the same class as the limewashes are the cement and cement-lime slurries, cement paints, and external distempers; these also require periodic renewal and are more suitable for use on the stronger grounds such as are provided by cement stabilized walls. Oil paints have been used in America, but the results of their tests are inconclusive and it seems very unlikely that oil paints would be satisfactory on earth walls in this country.

Whatever covering is used, the walls should first be allowed to dry out thoroughly, and should be brushed down to remove any loose material. A slightly rough surface has the advantage that it provides a useful key for washes and slurries, but it should be quite free from dusting.

RENDERINGS

Where a more durable covering is required, it is common practice to use renderings containing cement, but as these form a rigid external casing to the walls they should only be used with earth construction if special precautions are taken.

During the initial drying out of a cement rendering, shrinkage takes place which builds up stresses in the material. This state of strain tends to find relief by pulling the rendering away from the walls. As in the case of soft brickwork, the rendering is sometimes strong enough to pull away part of the outer surface of the earth walls, the process continuing as a result of unequal moisture movement of the walls and the rendering until the rendering breaks and falls away completely.

It is therefore necessary to avoid the use of too strong a mix and also to provide special means of bonding the rendering to the wall. There are two methods in common practice for doing this.

The first method is to fix chicken wire or expanded metal mesh to the walls with staples, or with wire ties threaded through holes left by shutter bolts, the holes being filled in afterwards with mortar. With block construction fixings for the mesh can be incorporated in the joints at the time of laying. The rendering is then applied to the mesh and forms a separate shell covering the walls. The chalk cottages at Hursley Park (Pls. 32 & 33) were rendered with a cement-lime rendering on chicken wire, and the rendering has remained in a very good condition.

The second method is cheaper, and has been found quite satisfactory for single-storey walls. After the shuttering is removed the wall surface is scored to form a key, and allowed to stand for several months to dry out. Just before the rendering is applied the walls should be sprayed with water to prevent the water in the rendering being absorbed by the walls. The undercoat is then laid straight on the walls and, while soft, is nailed to them with common nails. These are driven in flush with the undercoat, and are spaced at approximately 12-in. intervals from one another. A finishing coat is then applied. The nails should be as long as it is possible to drive in, and they should be random spaced so as to avoid defined planes along which the rendering might crack.

It is now considered better to allow renderings to cure without being dampened, but they should nevertheless be shaded from the hot sun to prevent them from drying out too quickly. The cement does not achieve quite such a high strength in this way, but the tendency to crack is lessened.

Rich mixes with smooth trowelled surfaces should be avoided, for they tend to crack badly, and whereas a smooth surface will throw off the rain readily, the fast-flowing sheet of water will enter rapidly where cracks occur. Cement-lime renderings with porous open-textured surfaces are more satisfactory, the finishes in order of efficiency being classified as roughcast or dry dash, pebble dash, scraped finish, and trowelled finish.

The following proportions for cement-lime mixes have been recommended for use with normal brick or concrete construction.[1] For normal exposure both undercoat and finishing coat

[1] R. Fitzmaurice, *Principles of Modern Building*. H.M.S.O.

may be of 1 part cement to 2 parts lime to 9 parts sand. For very exposed places a cement and sand undercoat and a stronger finishing coat in the proportion of, say, 1 part cement to 1 part lime to 6 parts sand may be required. The method of applying the rendering by throwing it against the wall which was used traditionally for lime roughcast, and which is common practice on the Continent today, is found in practice to give better results than a trowelled application.

The rendering should be discontinued at the plinth wall unless the damp-proof course be carried right through it and at the same time some form of elastic joint be provided to prevent horizontal cracking at the junction of the plinth and the earth wall. This cracking may otherwise develop from differences in moisture movement.

MUD PLASTERS

Apart from the ordinary cement and lime renderings, mud plasters called Dagga or Dogga plasters have been found to last successfully for two to three years in arid climates. A stiff mixture composed of 3 parts sand to 1 part clay is applied with a trowel. The plaster can be made more durable by stabilizing it with cement or other agent, just as the earth for the walls themselves is stabilized, and is then said to last for about three to four years or more.

TAR AND SAND

A well-established, but different method of waterproofing is to apply hot coal tar well blended with sand. When tar is available this is perhaps the best of all the treatments, as it is known to be efficient, and while only semi-permanent, is nevertheless relatively cheap to renew. Frequently, tarred walls were limewashed to avoid the dark colour, but this practice may lead to disappointment unless precautions are taken to prevent the tar from bleeding through. The main point to remember is that the limewash should not be applied for as long as possible—that is, several months at least after the walls have been tarred, and then only on a well-sanded surface.

BITUMINOUS WATERPROOFERS

There are now on the market many kinds of bituminous solutions, aqueous bituminous emulsions, and coal tar pitches suitable as waterproofers. The intending user should consult a reputable manufacturer as to suitable grades for use in the particular circumstances. The choice of treatment will depend on the degree of permanence required and the permissible cost. These waterproofers are available in different consistencies suitable for spraying, brushing, or trowelling on. Coloured emulsions are also obtainable for decorative finishes as an alternative to sanding and limewashing. Naturally, these waterproofers are more expensive than crude tar; and it should be noted that if once a tar or pitch or bituminous waterproofer be used, it is not easy to use any other type of treatment afterwards, because the bituminous base may bleed through or cause softening of oil paints, or prevent their drying.

SLATE HANGING, ETC.

For decorative purposes, or for exposed positions, there still remains the possibility of slate hanging and of weatherboarding. The fact that soil cement can easily be nailed into makes these methods a practical proposition. In a house in Kansas, flat fieldstones of Kansas limestone were placed against the outer side of the shuttering and soil cement was rammed behind them, the joints in the stone being afterwards pointed in mortar. This is another apparently successful method of obtaining a protective "facing."

INTERNAL FINISHES

As far as internal finishes are concerned, any of the treatments already mentioned can be used where suitable. It is worth noting that the internal face of earth walls used for cattle-sheds should be thoroughly protected by rendering or other means against urine, and also against licking by the animals.

Plastering has been successfully done on earth walls, but the weak nature of the ground should be borne in mind when deciding upon the type of plaster, and the walls must be allowed to dry out thoroughly before application. Obviously it would be possible to use wall boards on battens instead of plastering, and some time might then be saved in building, for it would not be necessary to wait until the walls were completely dry.

Houses with earth walls are themselves, today, something of a novelty to most of us in this country, and the adventurous will doubtless want to experiment also with methods of internal finishing which are not generally recognized practice. It is said, for instance, that earth walls treated with linseed oil and subsequently polished can be satisfactory and very agreeable, too. Other walls have been made with colouring matter included in a part of the mix which has been used as a plating to the surface. A Persian Gulf oxide in the proportion 1 : 60 is stated to give a good red, and carbon black in the proportion 1 : 300 has also been used.

It is possible, too, that trowelled finishes on soil cement walls could be papered direct. But that suggestion still awaits a trial.

CHAPTER SEVEN

Design

MANY of the design principles peculiar to earth building have already been discussed in the chapters dealing with the various types of walling, but no apology should be expected for repeating them here in summary form, for they are, after all, the basis of successful earth building and, indeed, of even more critical importance than were we considering normal construction.

The three most important principles have already been stressed in preceding chapters; two of them are summed up in the advice "Giv un a gude hat and a gude pair o' butes . . ." Add to this "*and* a good raincoat" and the three are complete.

EAVES OVERHANG AND GABLE ENDS

The "gude hat" implies in the first place wide eaves overhang on all sides to throw water clear of the walls. This protection is easily enough provided with a hipped roof, especially in one-storey buildings, but it must not be forgotten that with gable ends large areas of walling may be left exposed to the weather, unless special precautions are taken to protect them. Obviously there are several methods of overcoming this difficulty: drips, such as are found on old barns (Fig. 20), could be provided across gables at eaves level; rendering, tile or slate hanging, or weather-boarding could be used to give complete protection, or the gable ends could be built of an entirely different material. It is, in fact, somewhat difficult to build gable ends in earth unless blocks are used, on account of the short and diminishing layers which are involved. Whatever method be used, a widely overhanging roof covering will always give additional protection and will contribute, too, to the distinguishing characteristics proper to the earth-built house. The provision of a damp-proof course on the top of walls and under cills in order to safeguard the walls still further is sound practice, but is not absolutely essential where costs must be kept to a minimum.

Pl. 41.—Cob boundary walls at Ashwell, Hertfordshire.

Pl. 42.—Chalk boundary walls, Blewbury, Berkshire.

DESIGN

The analogy with the boots is more nearly accurate than might at first be supposed, for earth walls should have not only good

TRADITIONAL PROTECTION
OF GABLE ENDS.

WOOD

DRIPS

E.E.T.

Fig. 20.

foundations to support them, and an efficient damp-proof course to prevent moisture rising, both of which functions are required of a good sole, but they should also have a plinth wall of concrete, brick, or stone (Fig. 21) to protect the base of the wall from rain splashings and mechanical injury, not to mention the damage done by rats; here is where your boot "uppers" come in. The plinth wall should be about 12 in. high and should contain a damp-proof course; this should connect with the damp-proof membrane in the floor, if the floor be of "solid" construction.

DAMP-PROOF COURSE AND WATER TABLE

The damp-proof course may be of any recognized kind, but if the walls are to be "rammed," care should be taken to make sure that the damp-proof material is not fractured or punctured by ramming.

Chicken wire or broken glass is sometimes included at the base of the wall as a deterrent to rats, but should not generally be necessary if an adequate plinth wall is provided. A further safeguard against damage to the bottom of the wall is the provision of a water table; besides serving its special function such a water table looks well, and together with limewashed walls and a tarred plinth, combines to give the "earth cottage" its particular beauty.

WALLS

As for the walls themselves, the first essential is that they shall be thoroughly protected from rain and damp; this may be done in a number of ways, and the various coverings are discussed in Chapter VI.

With Stabilized Earth it is possible to build cavity walls with two layers of blocks or with thin leaves of *in situ* rammed earth. In this way the risk of damp penetration is minimized, but it is none the less necessary to have a protective covering to prevent attrition of the wall itself.

The simpler the plan-shape of the walls the better, for projections and angles involve difficulty in ramming and in fixing shuttering. It is simpler to build only one storey high, since greater protection is given by the roof, and if openings be carried up the full storey height, lintols, which sometimes give rise to trouble, may be eliminated. Moreover, it is easier to arrange for standard lengths of walling between openings, a procedure which simplifies shuttering. The widths between the openings should, of course, be wide enough to ensure stability, and this will depend on the general thickness of the walls.

The difficulty and expense of lifting earth into shuttering, or of pitching it on to Cob walls two storeys in height, is another reason for designing houses of one storey only. Furthermore, limewashes and other coverings are more easily applied and renewed, since long ladders and scaffolding are not required.

Corners should preferably be rounded or splayed so as to avoid spalling and injury. Stone or brick quoins are occasionally met with, but the bonding of stone or brick is difficult in practice, because of the different moisture movements of the two materials.

BONDING OF BRICKS, ETC., WITH EARTH WALLS

If it be necessary to bond bricks or stones with earth it is best to leave spaces in the earth wall to receive them at a later time,

DETAILS FOR THE BASE OF AN EARTH WALL

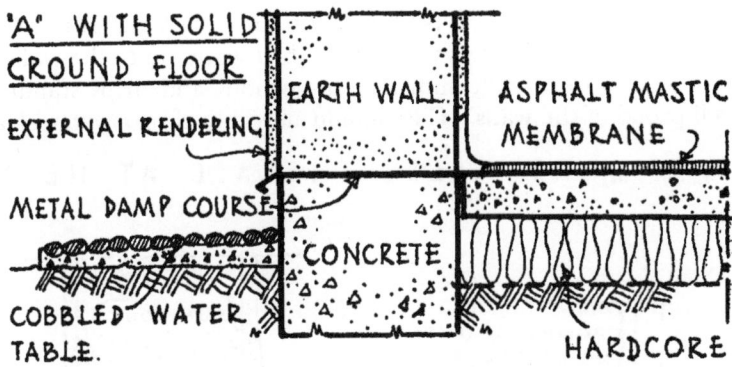

'A' WITH SOLID
GROUND FLOOR

EXTERNAL RENDERING

METAL DAMP COURSE

COBBLED WATER
TABLE.

EARTH WALL

ASPHALT MASTIC
MEMBRANE

CONCRETE

HARDCORE

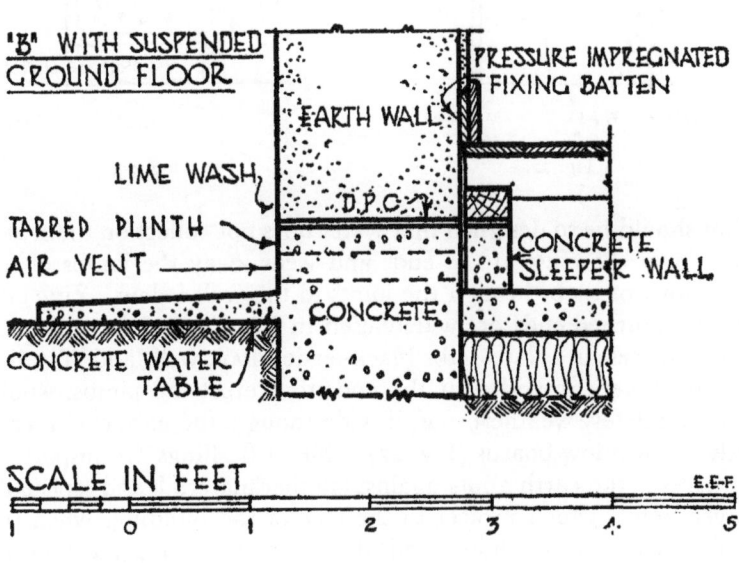

'B' WITH SUSPENDED
GROUND FLOOR

LIME WASH

TARRED PLINTH
AIR VENT

CONCRETE WATER
TABLE

EARTH WALL

PRESSURE IMPREGNATED
FIXING BATTEN

D.P.C.

CONCRETE
SLEEPER WALL

CONCRETE

SCALE IN FEET   E.E.F.

1   0   1   2   3   4   5

Fig. 21.

COB, PISÉ, AND STABILIZED EARTH

when some drying out will have taken place. In this way severe cracking may be avoided; but it is, nevertheless, just as well to avoid bonding earth and other materials by toothing, and to use instead continuous vertical rebates.

CILLS AND FLASHINGS

Just as with the roof it is necessary to provide a generous over-hang, so with all features which may shed water on to the walls is it necessary to provide sufficient projections and drips standing well proud of the walls. Cills should not only throw water clear

DETAIL AT CILL   DETAIL AT HEAD

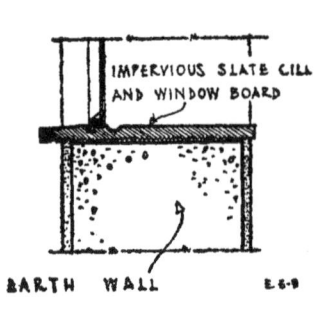

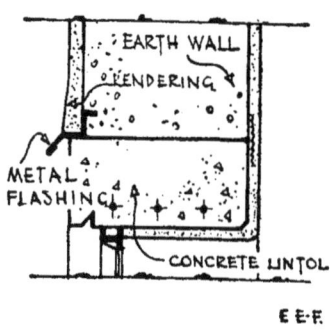

Fig. 22.   Fig. 23.

but should be so designed that water does not collect on them and run over the edge at the ends and wear away the jambs of the window, or seep through the joints to the wall below. Zinc cills with upturned ends are widely used in Europe and have proved very successful; slate slabs black-leaded look well, and if water grooves were provided at the ends to protect the jambs, would be completely weatherproof; if wide enough, the slabs could serve also as window boards (Fig. 22). Metal flashings are important wherever the earth abuts against another material, and especially over lintols which project to the face of the building, when the joint between the lintol and the earth has no protection by rendering (Fig. 23).

For the same reasons that drips and flashings are particularly necessary to shed water, it is also important to block out rain-

116

water pipes and other fixtures so that any water which may run down the outside of them does not come into contact with the wall. The illustration (Pl. 17) shows what may happen if one fails to observe this precaution.

OPENINGS

Openings should normally be spanned by wood or reinforced concrete lintols, which should have generous bearings of about 1 to 2 ft. on the walls at either side. They should extend the full width of the wall and must be supported on centres or temporary struts, especially in Pisé work, while building is still in progress, and until the walls have had time to harden a little.

Door and window frames are best placed on the outside of the walls, since the jambs are thereby made less vulnerable to the effect of dripping and rain splashing, and the necessity for providing a wide cill is avoided. From the point of view of interior design, windows placed on the outside of the wall have the merit of providing a wide, generous window board and deep reveals which, if splayed, reflect light inwards and are pleasing to look at. On the other hand there is some advantage in placing external door frames on the inside of a thick wall, since a porch is thus conveniently formed (Pl. 16).

FLOORS

Ground floors can be of "suspended" or "solid" construction and there are no special precautions to be taken in earth-walled houses which are not also applicable to normal brick construction. With suspended floors air bricks must be provided; with solid floors it is essential to provide a continuous damp-proof membrane of some kind which connects with the damp-proof course in the wall; this damp-proof membrane could itself be the floor finish if pitch or asphalt mastic were used. If, however, a wood surface be preferred, the usual precautions against dry rot should be taken; the wood should be treated with preservative, preferably under pressure, and the ends of the timbers should be kept out of contact with possible sources of dampness.

First floors should preferably be of timber joist construction. The joists should have a bearing on a concrete plate extending well into the wall, with or without the addition of a wood plate. There is some advantage in reducing the thickness of the wall at this point, so that the joists are not embedded in the wall. If, however, the wall is not reduced in thickness the earth should be cut away around the joists to provide some ventilation. The ends of the joists should always be protected by preservative, or by wrapping with a bituminous felt.

ROOFS

Roofs may be of normal construction, but it is important that there should be no thrust on the walls and that the rafters should

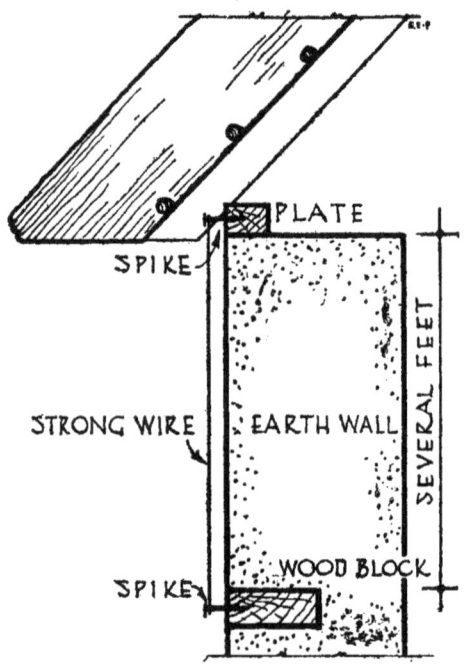

PLATE

SPIKE

STRONG WIRE

EARTH WALL

SEVERAL FEET

WOOD BLOCK

SPIKE

TRADITIONAL METHOD FOR TYING DOWN ROOF PLATE

Fig. 24.

be well tied. Clearly the roof members should rest on a wide plate, to distribute the load over the wall. Traditionally, the plate was sometimes tied down with wires as shown in the accompanying sketch (Fig. 24), and in contemporary practice is usually anchored to the wall by long bolts embedded in the earth wall.

When earth-walled houses are to be built in rural areas it might be economic to build the roof of unsawn timbers as shown in Mr. Thorpe's design (Fig. 10). These would probably necessitate a thatch covering to take up the unevenness of the timbers, but this would be wholly in keeping with the tradition of Pisé and Cob cottages, and would be an encouragement to keep alive an ancient craft.

PARTITIONS

Partitions might well be built of earth blocks if difficulty were experienced in otherwise constructing thin walls. Stronger

STUD PARTITIONS WITH EARTH FILLING

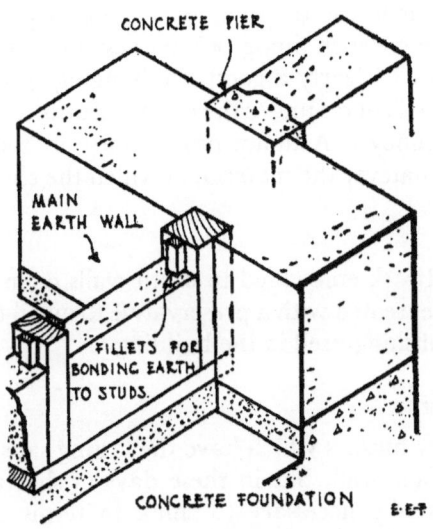

Fig. 25.

and thinner blocks than the normal could be made by using Stabilized Earth. The ends of the partitions should be tied into the main walls either by a continuous rebate in the main wall or by reinforcement, or by a wood fillet, or by ramming them integrally with the main wall. Stud and panel partitions are a further possibility, but care must be taken here also to bond the panel infilling to the studs. The easiest method is to fix fillets to the uprights (Fig. 25).

FIXINGS

Skirtings, picture rails, architraves, and other fixings can be fixed to plates or plugs embedded in the walls, or in some cases by nailing direct into the earth.

CHIMNEYS

Traditionally, chimneys were often built of earth, and examples remain where they have survived without defect. Nevertheless, it is probably easier to build them in brick or stone, and in so doing to group them centrally on plan, so as to avoid having to bond them to the external walls, involving, as that does, the risk of water penetration as a result of cracking. The best practice in chimney construction is to provide a lining of pre-cast fireclay liners: indeed, some such lining becomes essential if the chimneys be built of earth. Clearly a good overhanging coping and possibly a damp-proof course under the coping are especially necessary with earth chimneys. A damp-proof course at roof level is recommended, whatever the material of which the chimney is built.

WOODWORK

All the woodwork embedded in earth walls or in contact with them should be treated with a preservative to avoid the possibility of dry rot establishing itself in the building.

INTERIOR FINISHES

There are few finishes which have the lasting satisfaction of the traditional plaster wall, but in these days of shortage of skilled labour it is usually necessary to think in terms of alternative finishes. It is, perhaps, undesirable anyway to conceal completely the texture of the earth walling, since it can contribute much to the special character of the room, and methods of finishing, as for instance raw linseed oil brushed on to the earth and then polished, or a light limewash on earth blocks, have been suggested in Chapter VI with these considerations in mind.

CHAPTER EIGHT

A Successful Experiment

As one of their post-war land-settlement schemes in 1919, the Ministry of Agriculture and Fisheries decided to develop a considerable area owned by them on the outskirts of the village of Amesbury in Wiltshire, so as to provide a number of small-holdings.

Some thirty-three houses were erected. Most of them were of normal construction, but the design and construction of others was specifically experimental. All the houses were built by the Ministry by direct labour, but in the case of five of them the designs and the supervision of the building operations were undertaken by officers of the Department of Scientific and Industrial Research.

At the time, a number of suggestions had been made to the Research Department as to the use of unorthodox building materials. The difficulties and cost of transporting materials to rural areas, and the dearth of skilled labour, clearly suggested then, as now, that the use of local materials and of local traditional building techniques might be of value in carrying out the post-war building programme. The project for the small-holdings provided an opportunity to study such traditional methods of building and, with the addition of certain improvements in design and technique, to assess their value in terms of contemporary standards.

As the site chosen was at Amesbury it was natural that amongst the building methods which suggested themselves for experiment should be the old earth-walling methods. Amesbury and the adjoining villages had, in fact, preserved a lively tradition of Chalk-mud walling which was very similar to that of Cob, practised farther west in Devon.

Chalk was readily available on the site, and although for the sake of completing the experiment an attempt was made to import suitable earth for walling in Cob, the difficulties at the time proved insuperable and the experiment was ultimately confined to variants in chalk walling.

COB, PISÉ, AND STABILIZED EARTH

Of the five cottages whose erection was supervised by officers of the Department of Scientific and Industrial Research, three were of earth construction, one was of brick, and one of concrete; of the remaining houses in the Ministry's scheme a further six were of earth. These included a single cottage with walls of Chalk Mud built in the old traditional method. Another single cottage and a pair of semi-detached cottages had walls of chalk Pisé, while a further single cottage was built of chalk and cement blocks that had been previously made on the site in a "Dricrete" machine. This last has proved one of the best.

It is proposed to discuss in greater detail the three earth-walled houses supervised by officers of the Department of Scientific and Industrial Research, which will be referred to as No. 1 Ratfyn Lane, No. 2 Ratfyn Lane (originally called No. 4 Ratfyn and No. 5 Ratfyn), and No. 10 Holders Road,[1] and in addition No. 13 Holders Road. The walls of No. 1 Ratfyn Lane were built of chalk and straw, those of No. 2 of rammed chalk with an admixture of chalky gravel, and those of No. 10 Holders Road of rammed chalk stabilized with cement.

In an enquiry made at the time, many of the villages around Salisbury and Amesbury, and in the district about Exeter in Devonshire, were visited. In the Amesbury district a number of examples of chalk construction were found. Amongst other typical features were cut chalk-rock arches over windows, with quoins to windows, door jambs, external angles, and chimney stacks in deep red bricks, wide overhanging eaves formed by the generous thatch of the roofs, and plinths of brick or stone, sometimes carried up to ground-floor window level. Window and door openings were generally spanned with an oak lintol which often proved inadequate to support the weight of the walling above. Many of the chalk walls, whether of hewn chalk rock or of Chalk Mud with external rendering, were seen to have disintegrated at the foot of the wall. This was, no doubt, due to the omission of damp-proof courses and to the additional exposure at that point. It was significant that the upper part of the walls which was protected by the eaves had not generally so suffered.

In the district round Exeter, the traditional characteristics were much the same as those in Wiltshire. Most of the Devon-

---

[1] Report of the Department of Scientific and Industrial Research. *Experimental Cottages*, by W. R. Jaggard. H.M. Stationery Office, London, 1921.

shire cottages were plastered externally, but brick was not so commonly used for door and window quoins as in Wiltshire. The roofs were usually of thatch.

In Wiltshire, it was noticed that the chimney stacks were invariably built of brick, and very few were placed on external walls. In no single instance was a chimney stack found constructed of chalk. In Devonshire, on the other hand, external chimneys were quite common and the lower portion was very often built of Cob, forming a large open-hearth fireplace internally, with an oven at the side. The upper part of the stack was usually finished with a few oversailing brick courses.

The experimental cottages were designed to include some of these characteristics, and an endeavour was made to obtain the services of local tradesmen well versed in the art of "Mud Walling." At the same time a number of modern improvements, such as damp-proof courses, were introduced, to make the houses more weatherproof and more durable.

PLAN

The plans of the D.S.I.R. cottages were based on a standard type plan (Fig. 26), but departures from this were made in all three earth-walled cottages for various experimental purposes which will be apparent in the detailed descriptions which follow.

FOUNDATIONS AND DAMP-PROOF COURSES

In each case the foundations were taken up to the level of the chalk and consisted of a bed of concrete poured into trenches without boarding.

No. 10 Holders Road.—Fig. 25 shows the foundations to cottage No. 10 Holders Road. The damp-proof course was of two courses of slates. The walling was then continued to a height of 3 ft. 6 in. above the ground level by building inner and outer walls 4½ in. thick in English garden-wall bond, the 9-in. space between these walls being filled in with almost dry well-rammed chalk. These walls were necessarily 18 in. thick in order to take the chalk and cement walling over, and the chalk filling was adopted in order to reduce the cost. The main internal 9-in. walls were carried down to the chalk and rested upon a concrete 1 ft. 1½ in. wide and 1 ft. 6 in. deep.

COB, PISÉ, AND STABILIZED EARTH

No. 1 Ratfyn Lane.—In cottage No. 1 Ratfyn Lane there was a 9-in. plinth of brick faced with selected but unsnapped flints, with brick quoins. The damp-proof course was of two courses of slates. Fig. 26 shows a section through these foundation

MINISTRY OF AGRICULTURE TYPE PLAN 'C'

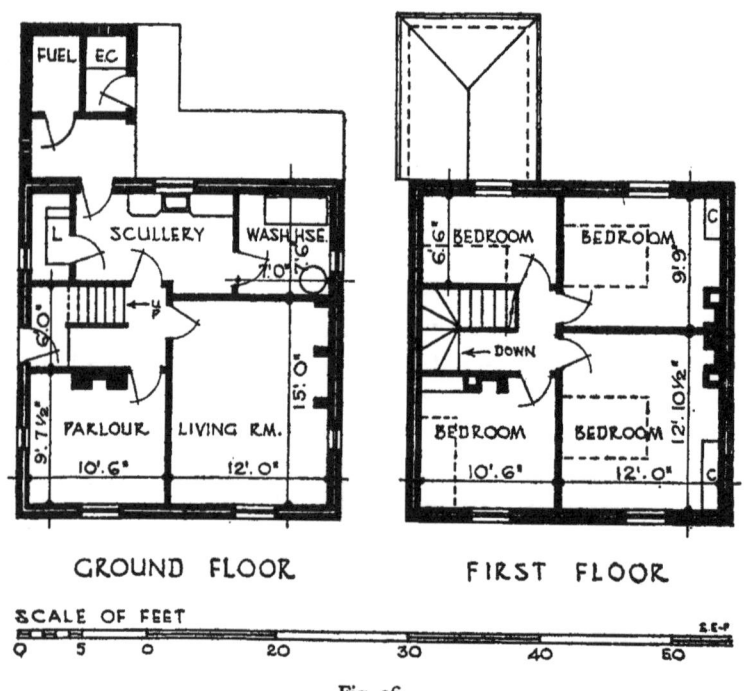

GROUND FLOOR   FIRST FLOOR

Fig. 26.

walls, which are characteristic of much of the old cottage work in the district.

No. 2 Ratfyn Lane.—In cottage No. 2 Ratfyn Lane the foundations and plinth wall were of concrete in the proportions 7 of Amesbury gravel to 1 of Portland cement. The damp-proof course consisted of two coats of hot coal tar well sprinkled with sand (Fig. 27). Some difficulty was experienced at the time in preventing the tar from running, but the method proved efficient and remarkably inexpensive, and no damp has since appeared above the damp-proof course.

SHUTTERING AND WALLING

The traditional methods of building up Chalk Mud and Cob walling did not commend themselves because of the time required for drying out and the risk of shrinkage cracking; they seemed also to be unnecessarily clumsy. It was decided that better results would be obtained by the use of shuttering, and of a drier earth mix than was possible with either Chalk Mud or Cob. It therefore became necessary to design some form of standard shuttering, which should be of simple construction, such as could be made locally. It was likewise necessary to devise something that could be easily handled, readily struck, and reassembled with little or no skilled labour, and of such a kind as could be adapted for any position on the walls of a building.

The set of standard shutters which was used for the external walling in No. 1 Ratfyn Lane and No. 10 Holders Road is illustrated in the drawing (Fig. 27).

The shutters for cottage No. 10 were made so that each layer of the walling was approximately 18 in. deep, but it was thought desirable to obtain a greater depth, and an additional board was eventually added at the top. At the same time, the wires shown in the drawing were eliminated, because it was found difficult to ram between them. Hardwood wedge-shaped cross bearers near the bottom of the shutters were substituted for the wires. In 1921 the approximate cost of a set of shuttering consisting of one corner section and two side sections was £25.

No. 1 Ratfyn Lane.—The earth for cottage No. 1 Ratfyn Lane consisted of the chalk which lay directly beneath the subsoil. The larger lumps of chalk were broken to about a 2-in. gauge and the chalk and straw were mixed together on a platform with just sufficient water to make the mixture plastic when worked up in the hand. The average moisture content was about 8 per cent. of the weight of the chalk.

The mixture was thrown into the shutters in layers about 3 in. deep and was at first rammed with flat rammers. It is specially noteworthy that this shape of rammer was found in practice to "case harden" the surface of the layer which is being rammed.

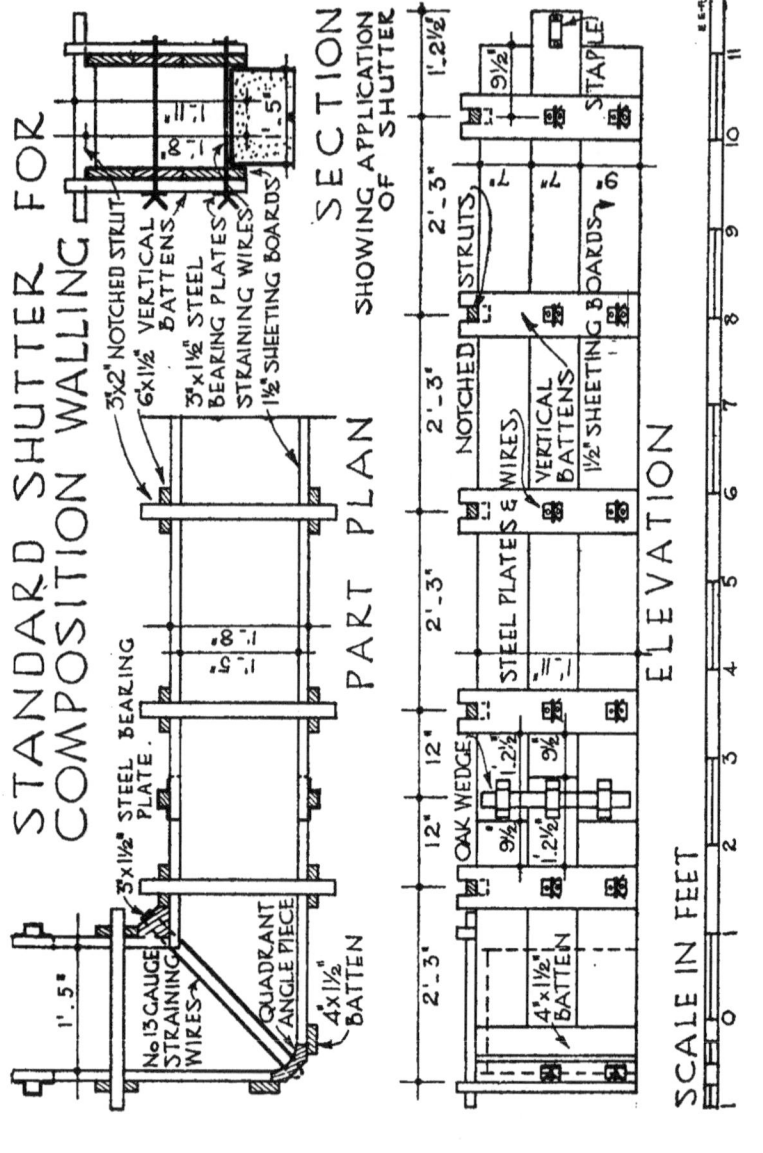

STANDARD SHUTTER FOR COMPOSITION WALLING

3½x2" NOTCHED STRUT
6x1½" VERTICAL BATTENS
3x1½" STEEL BEARING PLATES
STRAINING WIRES
1½" SHEETING BOARDS

SECTION

SHOWING APPLICATION OF SHUTTER

3x1½" STEEL BEARING PLATE.

No.13 GAUGE STRAINING WIRES

QUADRANT ANGLE PIECE

4"x1½" BATTEN

PART PLAN

1'.5"

1'.8"
1'.5"

STAPLE
9½"
1'.2½"

2'-3"

NOTCHED STRUTS

7"
7"
6"

2'-3"

2'-3"

STEEL PLATES & WIRES

VERTICAL BATTENS

1½" SHEETING BOARDS

1'.11"

12" 12"

OAK WEDGE

1'.2½"
9½"
9½"
1'.2½"

4"x1½" BATTEN

2'-3"

ELEVATION

0 1 2 3 4 5 6 7 8 9 10 11

SCALE IN FEET

Fig. 27.

126

Wedge and heart-shaped tools, as shown in Fig. 28, were subsequently used and gave much better results.

Projections in the form of external chimney stacks (Pl. 43), and of a string course at the level of the upper-floor window cills were

RAMMERS   FOR   PISE   WORK

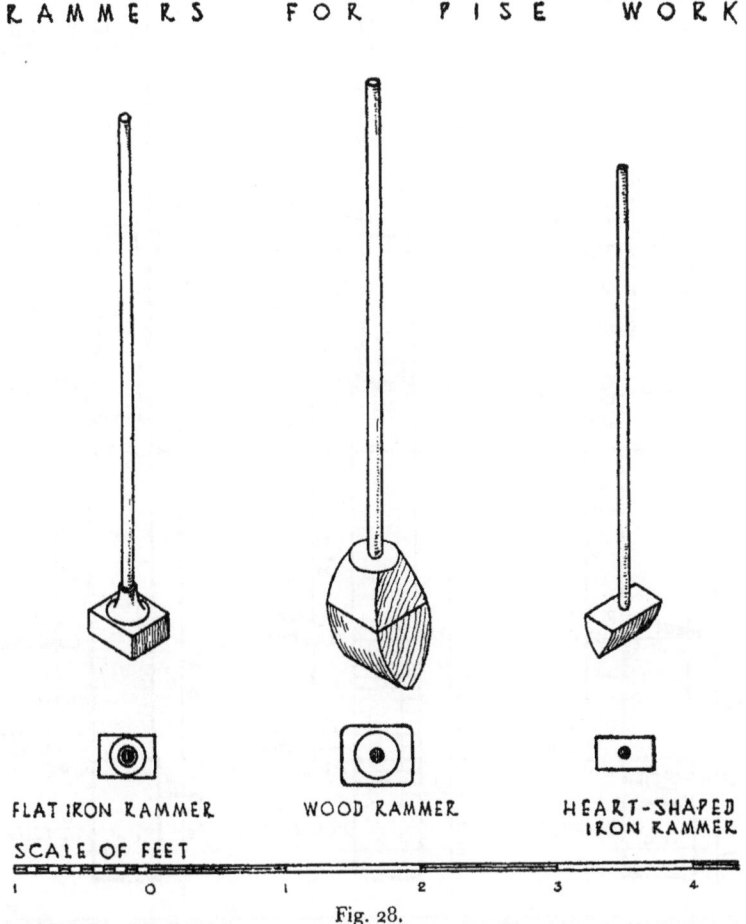

FLAT IRON RAMMER   WOOD RAMMER   HEART-SHAPED
                                     IRON RAMMER

SCALE OF FEET

1   0   1   2   3   4

Fig. 28.

incorporated with the express purpose of investigating the adaptability of the standard form of shuttering to the formation of such projections. The difficulties which resulted from this experiment made it clear that when shuttering is used it is very necessary to avoid projections wherever possible.

SECTIONS·THROUGH·MAIN·EXTERNAL·WALLS·OF·COTTAGES

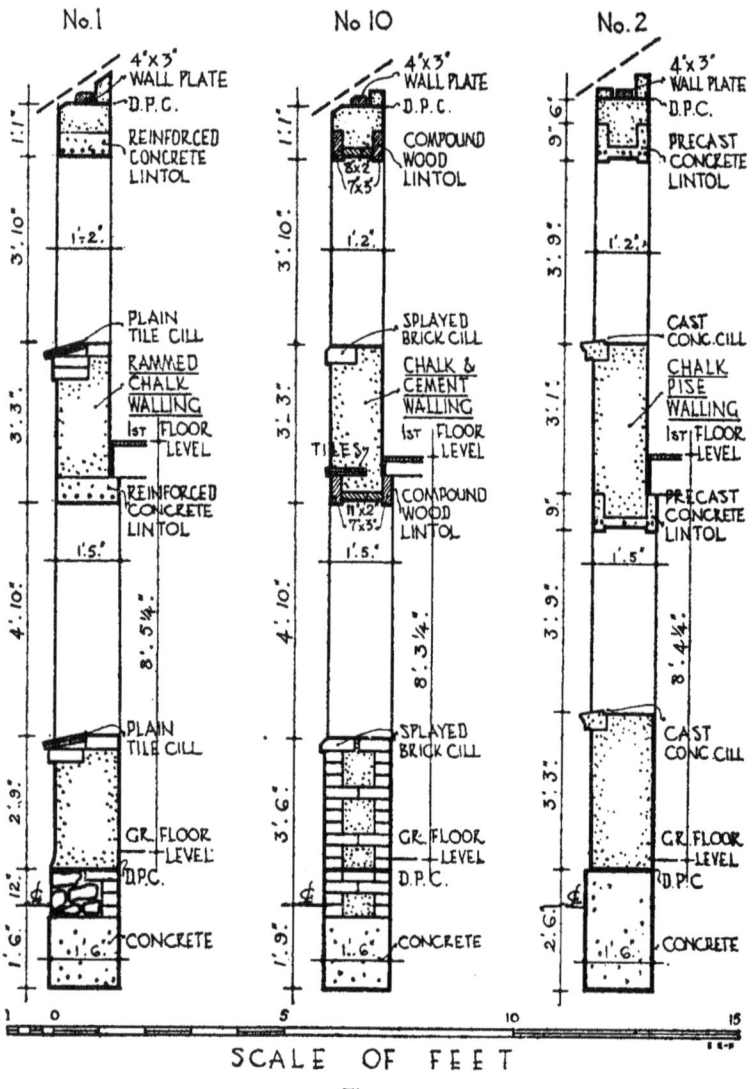

Fig. 29.

The section shown in Fig. 29 shows the varying thicknesses of the main wall and some of the constructional details. The 6 in. by 3 in. concrete wall plate was cast *in situ* in the shutters.

Pl. 43.—No. 1 Ratfyn Lane.

Pl. 44.—No. 2 Ratfyn Lane.

Pl. 45.—Projecting chimney stack and porch at No. 1 Ratfyn Lane.

Pl. 46.—Attrition at a window cill due to the outer tiles of the cill not being tilted inwards to throw the water off the front edge.

Pl. 47.—Exposed walling on an outhouse at No. 10 Holders Road.

The fireplace openings were formed in chalk and were subsequently lined with brickwork. The flues consisted of 9-in. circular field drains with butt joints and were packed round with rough concrete to prevent calcination of the chalk.

The walling of this particular cottage was carried out during the winter, and special precautions were necessary to protect it from the weather. In order to avoid disintegration from rain and frost, long straw was bound to battens and hung along the tops of the walls. Strips of asphalted felt were also used as a protection against rain. Nevertheless, the view was held that the actual construction of the walls would present little or no difficulty during the fine days of spring and summer. Very hot days should be avoided for building, because the chalk has a tendency to dry out too quickly. Possibly the best months of the year for putting up this type of wall would be March, April, May, and June.

When No. 1 Ratfyn Lane was examined in 1927, and subsequently in 1945, the walls were seen to have cracked fairly extensively on the outside, but in only two places had the cracks penetrated to the inside plaster, and there was no evidence of leakage through the walls. The external rendering of lime plaster and the internal lime plaster finish had adhered well, and both were satisfactory.

The low rate of warming up of these earth walls tends to accentuate condensation when intermittent heating is used, and it was evident that condensation did, in fact, sometimes occur. The difficulty could easily be overcome, as in No. 2 Ratfyn Lane, by including a suitable lining of low heat capacity, as for instance a fibre board, which warms up rapidly.

No. 2 Ratfyn Lane.—An attempt was made to build No. 2 Ratfyn Lane in true Devonshire Cob, but it was not possible to obtain the right material, and the walls were finally built with a graded earth mixture of chalky gravel and fine soil by the normal Pisé method. The earth was rammed between shutters in a relatively dry state and without the addition of straw.

There were indications at this early date that moisture content was an important factor both in the process of ramming and in the degree of ultimate cohesion of the soil. As this wall was built in fine weather, the main difficulty encountered was the too rapid drying of the earth during storage before use.

The masonry for the offsets to the chimney stacks and for the slightly projecting stone porch was not built in during the ramming of the walls, but was inserted later into spaces left to receive it. In this way it was thought that shrinkage cracks between earth and masonry might be avoided. This was in fact successful.

The wall plate for the upper floor was of deal, 6 in. by 2 in., previously creosoted, and laid flat, and the roof wall plate was of 4 in. by 3 in. deal and rested upon the centre of the wall on a felt damp-proof course.

In contrast to the walling of No. 1 Ratfyn Lane, which was undertaken in the winter and took five months, the walls of No. 2 Ratfyn Lane were built during more suitable months, namely between April and June, and took only ten weeks.

There has been no evidence of dampness in this cottage and there are still very few cracks in the walls after twenty-five years. The earth mix was considerably drier than in the case of No. 1 Ratfyn Lane, and as might be expected the cracking was correspondingly less. The walls were originally covered externally with a lime slurry, but have now a cement roughcast. This roughcast has come away from the Pisé backing, although it is quite sound in itself. This is likely to happen with such renderings, unless a good mechanical key be provided in the form of reinforcement pegged to the wall. An inspection by P. W. Barnett, of the Building Research Station, made in 1927 before the existing rendering was applied, showed the walls remarkably free from cracks. This is undoubtedly to be associated with the low water content of the mix and is a valuable piece of evidence.

No. 10 *Holders Road.*—Cottage No. 10 Holders Road was built in a stabilized soil mixture, composed of chalk and Portland cement. The chalk was similar to that used in traditional Chalk Mud walling.

The following is an extract from the specification for the walling:

"The chalk, after digging, will be broken to pass a $1\frac{1}{2}$-in. mesh. It will then be mixed dry, on a boarded platform, with $\frac{1}{20}$th of its weight of Portland cement, by turning over at least three times, as in concrete mixing, and none of the mixture must be allowed to stand for more than thirty minutes before using."

It was not found necessary to add any water to the mix, as the moisture content of the chalk averaged 20 per cent. throughout the work.

A SUCCESSFUL EXPERIMENT

As soon as the earth was mixed it was placed in the standard shutters in layers of about 2 to 3 in. and thoroughly rammed with wedge-shaped rammers. These, together with other wooden rammers, were abandoned in favour of heart-shaped metal rammers when it was found that the material adhered to them unduly (Fig. 28).

The main external walls were 1 ft. 5 in. thick on the ground floor and rested upon the brick and chalk base previously described. The upper walls were reduced to 1 ft. 2 in. thick and a bearing for the wall plates for the upper floor was provided by header bricks which were built in at 18-in. centres with the ends flush with the inner surface of the lower wall.

The shutters used gave layers of walling of a depth of about 20 in. On top of the second layer above the brick base, a strip of $1\frac{1}{2}$-in.-mesh wire netting was laid in as long lengths as possible, and overlapping each other at the angles. The straining wires of the shutters rested upon the netting, and kept it in position until the chalk walling was placed. Similar strips of wire netting were placed at every alternate course, i.e. at 3 ft. 4 in. intervals in the height. The object of the wire netting was to provide a horizontal reinforcement.

The flues were built of 9-in. diameter field drains, and the whole of the chimney stack and external breast was carried up in stabilized chalk, and was finished at the top with a 9-in. concrete cap, and short terra-cotta chimney pots.

This early test use of a Portland cement mix had as its object the providing more resistance to attrition and better weather protection. The illustration (Pl. 46) shows a small part of the walling on an outhouse which has been exposed for twenty-five years, together with a part of the plastered main wall of the house. Experience shows that the unprotected outhouse wall has not, in fact, suffered from attrition. The main walls of the house, however, were found to have cracked extensively when they were examined in 1927. The wire-netting reinforcement did not appear to have prevented initial shrinkage cracking. These cracks in the chalk walling are no worse today. They are probably to be associated with the use of cement and also probably with a wet mix, and this result should be compared with that of the walling of No. 2 Ratfyn Lane noted above. The majority of the cracks have occurred largely between heads and cills, as is common also in concrete buildings. Nevertheless, once the cracks have formed, they can be sealed and will not then give trouble. A subsidence crack noted in the plinth wall suggests that the foundations, which were designed without projecting footings, were insufficient for the load, despite the chalk bearing.

No. 13 Holders Road.—This cottage, which was not under the supervision of the Department of Scientific and Industrial Research, was, nevertheless, of an experimental nature, and has proved to be one of the most successful.

The walls were built in two leaves of chalk and cement blocks of the proportion of 12 parts of chalk to 1 part cement, with a 2-in. cavity between. The blocks, which were 4 in. wide, were made in a "Dricrete" machine, and were laid in the wall with galvanized wall-ties inserted in the usual manner for cavity walls. Externally the blocks were rendered with a lime slurry and internally they were finished with a lime plaster.

Two main advantages in the use of block walling as compared with monolithic walling are demonstrated in this cottage: first, that with blocks it is considerably easier to build a cavity wall, whose obvious merits there is no need to discuss here; and second, that most of the shrinkage takes place in the blocks while curing, and so minimizes cracking of the walls. The walls of No. 13 Holders Road are, in fact, remarkably free from cracks.

Obviously, in cavity earth walls particular care must be taken to provide cavity gutters and to ensure that the inner leaf is strong enough to take the loads required of it, or that these are distributed across the two leaves.

Some of the other experimental features in these earth-walled cottages are worth noting.

FLOORS

In No. 1 Ratfyn Lane the floors of the scullery larder and outbuilding were of red brick laid flat in 2 : 1 lime mortar on a rammed chalk filling. The remainder of the floors were covered with a 1½-in. layer of 8 : 1 concrete finished with a ½-in. thickness of cement and sand. Although carpeted, the concrete floor does not provide sufficient heat insulation, and it is also generally recognized today that a damp-proof membrane is essential, or at least a sufficient underlayer of dry hard-core.

In No. 2 Ratfyn Lane the under floors of the living-room and parlour were composed of a layer of rammed chalk filling on which was placed a 2-in. layer of well-rolled tar paving, composed of graded granite chippings thoroughly impregnated with hot coal tar and dusted with fine sand. The flooring was of 1-in. tongued and grooved boarding nailed to 1 in. by $\frac{3}{4}$ in. wood strips bedded flush with the tar paving. This method of flooring seems to have been satisfactory, and has not given rise to trouble during its life, and there have been no signs of dry rot. It would, of course, be advisable to impregnate the wood strips and to brush treat the undersides of the tongued and grooved boarding with a suitable preservative. The more normal practice today is to incorporate a damp-proof membrane sandwiched in a concrete base and to lay the floor surface above this membrane.

CILLS AND LINTOLS

Cottage No. 1 Ratfyn Lane has cills of two courses of plain roofing tiles as shown in Fig. 30. These were bedded in 3 : 1 cement mortar and overlapped the window opening by 3 in. The photograph (Pl. 45) shows the effect of attrition due to dripping of rain water from the edges of the cills at the jambs. In order to avoid this, the cill should have ends tilted, or be so constructed as to direct the water to the front edge only.

The lintols of this cottage were of reinforced concrete, rectangular in shape, and were covered with the external rendering. Their outline can be traced, but beneath the rendering there is no serious cracking and no water has penetrated the wall.

Cills and lintols for No. 2 Ratfyn Lane were of pre-cast concrete (the cill is shown in the drawing, Fig. 30), and have proved satisfactory. The cills, in this case, were placed in position after the completion of the walling. An important feature of the lintol design is the projecting lead apron.

In No. 10 Holders Road the cills were formed of splayed edge bricks set flat in cement mortar, with an oak cill placed on top without a water bar. Moisture has penetrated at the joint of the oak cill with the bricks and also through the wood lintols. Brick cills of this kind are, in any case, not to be recommended unless a damp-proof course is first laid on the wall.

The defects in the design of the cills and lintols in this cottage show how important it is, particularly in earth-wall construction, to avoid straight-through joints and to provide a liberal arrange-

133

ment of aprons, throats, and drips to throw the water clear of the walls. The methods of protecting the walls from rain and its effects are probably the most important elements of design in earth-wall construction; there is already sufficient evidence to

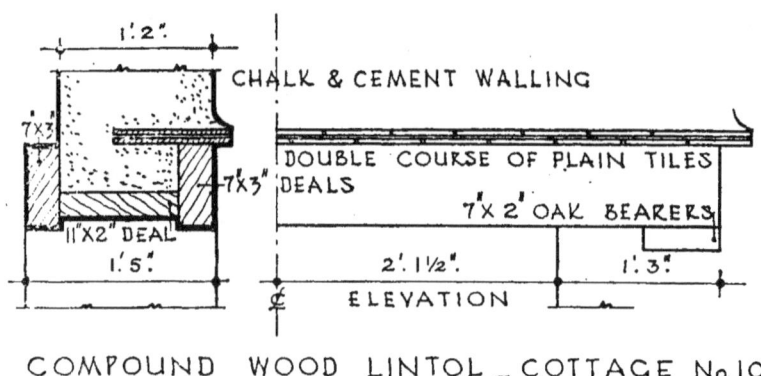

COMPOUND WOOD LINTOL _ COTTAGE No.10.

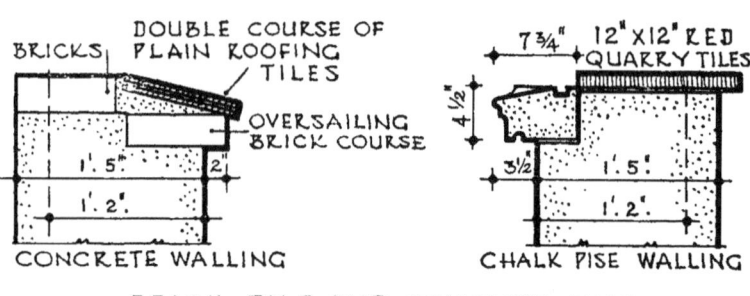

BRICK, TILE, AND CONCRETE CILLS

FOR COTTAGE No.1.   FOR COTTAGE No.2.

SCALE OF FEET

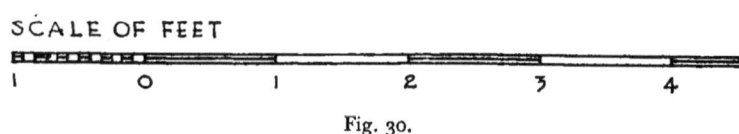

Fig. 30.

show that even with a relatively primitive technique and little knowledge of soil mixes, earth-wall construction can produce a permanent and satisfactory dwelling, provided the design is truly functional and employs the proper means of eliminating the effects of rain and dampness.

A SUCCESSFUL EXPERIMENT

The value of the experiment at Amesbury is that we have evidence of the behaviour of walls in cottages now twenty-five years old, and that we can study in some detail the records of the methods of building which were employed. Since the soil mixing and other building processes were carried out under controlled conditions, it is possible to make fairly accurate comparisons of the advantages of one type of walling over another, and to predict with reasonable certainty what results will be obtained by following any one of the methods described.

CHAPTER NINE

Some Practices Abroad

A s we have seen, building in earth is traditional to many parts of the world, and has been practised for as long as history is recorded. The Great Wall of China is itself testimony to the permanence of rammed earth, and the earthen ramparts of Kano, from which the cannon balls of a besieging army are said to have rebounded, testify to its strength. There are, indeed, many historic examples of earth-walled buildings of all kinds: among them native huts and peasant dwelling-houses, churches, Emirs' palaces and Chinese pagodas; but what is of special interest is the extent to which earth walling has been used in modern times, both as an alternative to modern methods and materials, and as a development of native traditions.

In many countries the traditions of earth building have been kept alive both by local practice and also by published information. For instance, in 1911 the Department of Agriculture, New South Wales, issued a *Farmer's Handbook*, containing information on Pisé and Adobe; in 1920, a number of German pamphlets were issued on the subject of earth building, and the Report on the Experimental Houses at Amesbury was also published; the first edition of this book appeared also in that year; from time to time P.W.D. Manuals have given recommendations for various kinds of earth buildings, and recently in the United States of America a number of bulletins and articles have been published containing records of experimental buildings and of laboratory tests.

Between the wars Pisé was practised in India and Malaya, and many parts of the colonial empire, and was found generally to be a cheap method of building which could be carried out satisfactorily with unskilled labour, provided that there was adequate supervision.

It is evident from the published literature on the subject that Pisé and Adobe building have made a contribution to the development of housing and farm settlement schemes in parts of the United States of America. Several hundred Pisé houses have been erected in Puerto Rico by the Puerto Rico Reconstruction Administration; the Bureau of Indian Affairs has erected Pisé houses on the Indian reservations in South Dakota, and there has been further work in Alabama by the Resettlement Administration. Private individuals have built Pisé houses for themselves, and have paid high tribute to them. One such owner built a one-storey farmhouse and a large garage in Pisé on his ranch at Hettinger, North Dakota, and was besieged by so many visitors wishing to obtain information about rammed earth walls, that he found it expedient to erect a large notice board on his premises with photographs of all the building processes and instructions on the method of building. He is said to have remarked, "It is the best home I ever lived in."

During this last war a number of military camps and other wartime buildings were erected in Pisé in various parts of the colonial empire; but perhaps the most striking recognition of the value of the use of earth walling was the issue, in January 1942, of an official order by the German Minister of Labour (Reichsarbeitsminister) about emergency housing in Germany's newly acquired eastern provinces. Full instructions were given, and a special advisory bureau and training centre (Lehr- und Beratungstelle Lehmbau) was instituted at Posen, to which men were sent for instruction.

GERMANY

IN 1920 there was a revival of the tradition of earth building in Germany, and a number of fairly large buildings were erected, some of them of two and three storeys. During this last war there was a further revival. The following are extracts translated from the instructions mentioned above.[1]

EARTH FOR BUILDING

Earth fit for building is a naturally occurring mixture of clay, sand, and gravel. A too clayey earth should be tempered with sand or gravel, or with fibrous material such as straw, heather, and so forth. A too sandy soil should have an admixture of clay.

Before deciding upon a suitable soil mix and method of preparation, the building authority should seek an official

---

[1] "Haüser und Ställe in den Ostgebieten als Lehmbauten," *Bauwelt*, February 27th, 1942, Vol. 33, No. 9/10.

report on the soil. (This report was to be had from the Advisory Bureau in Posen.)

BUILDING SEASON

Work on earth building is usually limited to the months from May to September, but in districts where the weather is particularly favourable may extend from April to October. Pisé and Cob walls should be finished by the middle of September.

The earth should normally be dug out and piled loosely several months before starting to build—that is, some time in the autumn of the previous year. Both Adobe blocks and earth bricks may be stored for some time if they be well protected against rain and frost.

SUPERVISION

Earth building should be carried out only under the supervision of a thoroughly experienced craftsman, who must be prepared to prove his qualifications.

EXECUTION OF THE WORK

Adobe and earth bricks should be protected from ground moisture and from rain during storage on site before being laid.

During the building and the drying-out time the work should be thoroughly protected against heavy rain on the top by tarpaulins or other suitable materials, and on the sides by reed or straw matting, or wooden planks, etc., unless the roof is already constructed on either temporary or permanent supports. Internal partitions should not as a rule be built until protection is afforded by the rest of the building.

CHOICE OF TECHNIQUE

The choice of technique will depend upon the type of earth to hand, and upon the available building labour and organization, as well as upon the extent of the work to be done.

THE VARIOUS BUILDING TECHNIQUES PERMISSIBLE

The approved methods described in pages 139 to 149 are permissible under the conditions specified above, provided that they suit also local conditions and provided they are built in accordance with the technical rules peculiar to earth building. Other earth-building methods should be used only if they are first adequately tested.

Pisé (Lehmstampfwände)

Gravelly earth is the best for Pisé (Steinreiche Berg oder Gehängelehm), and some sort of fibrous material should be added. The dampened ingredients should be uniformly mixed. If hard materials such as pebbles, broken stones, tile fragments, or breeze are rammed in to provide a better key for the rendering, they should be placed evenly on both sides of the wall. The mixture should be placed in equal layers of 10 to 15 cm. and uniformly rammed. At the level of the window cills and lintols reinforcement in the form of laths, small branches, withes, and so on, should be rammed in.

The thickness of load-bearing walls should be at least 40 cm., and that of internal partitions 30 cm.

Cob Walls (Wellerwände)

For Cob walls the best earth is one containing more clay to which straw, in lengths of 40 to 50 cm., has been added and well mixed in by stamping. The walls are built up in several courses, each not higher than one metre, the earth being pitched on with strong forks. The earth is allowed to project about 10 cm. on either side of the plinth wall, and after setting, the walls are pared down vertically.

The finished thickness of the walls should be at least 45 cm.

Pisé Blocks (Quaderwände)

Pisé blocks should be rammed in single moulds from the same damp earth as is used for Pisé. The blocks should be laid by skilled labour with a thin earth mortar and should be properly bonded. The joints should be as thin as possible, and should be raked out on the outside. Hollow blocks are not permissible. The most convenient size for blocks is 40 by 25 by 15 cm. (Einmann-quader).

The external walls should be at least 40 cm. thick and weight-bearing partitions at least 25 cm. thick, unless greater thicknesses be required for heat insulation.

COB, PISÉ, AND STABILIZED EARTH

ADOBE BLOCKS (PATZENWÄNDE)

Adobe blocks are made by tamping earth of a plastic consistency into wood moulds—usually a multiple of the standard brick size (Reichsziegalformat)—and after being dried are laid in the same way as Pisé blocks.

External walls should be at least 38 cm. thick, and weight-bearing partitions 25 cm. thick.

STUDDED WALLS (LEHMSTÄNDERWÄNDE)

Studded walls can be designed so that the weight of the ceiling and roof is taken by wooden supports. The earth cladding should be thick enough to give sufficient thermal insulation.

It is also possible to design the walls in such a way that a few widely spaced wooden studs are used merely as temporary supports for the roof while the earth walls are being built; when the walls are finished, the load is transferred from the supports to the walls. With this system a thickness of 40 cm. is sufficient for the external earth walls.

This method is known as Dünner-Verfahren. Mud bricks, which are usually made in an extruding machine, are laid without mortar whilst still in a plastic condition, and under the protection of the roof; the supports are left in the wall and are later relieved of their load. Care must be taken when laying the bricks that there are no spaces right through the wall between the bricks. Wooden reinforcement as described for Pisé should be used.

WALLS FOR FOUNDATIONS AND BASEMENTS

Earth as a building material should not be used below ground level. Basement and plinth walls built of stone or concrete should be carried up to a height of at least 50 cm. above the surrounding ground level in order to protect the walls from splashings.

WALLS

With the exception of attic walls and gable ends, external walls of earth may only be built one storey high.

For protection against ground moisture a damp-proof course should be laid on the plinth or basement wall and should project through the external rendering. A course of bricks or a layer of concrete should be laid on top of the damp-proof course to a height of at least 5 cm. above the level of the ground floor.

For protection against moisture penetration from above, a course of bricks in cement-lime mortar, or a damp-proof course laid on a suitably hard bed, should be provided.

The external surfaces of earth walls should be protected by a durable waterproof covering. This covering should be either an impervious rendering applied as described below, or some cladding, such as weatherboarding, tiles, shingles, or slates. Gable ends in earth should in any case always be protected by one of the last-named claddings.

Projecting plinths, cornices, window architraves, and so forth should be avoided. Damp-proof courses should be laid below window cills to protect the wall beneath.

When thin attic walls are placed on thicker walls beneath, eccentric loading should be avoided. Wall plates should be as wide as possible and should be laid flat like planks, so that the pressure is equally distributed on the wall.

Party walls can be built of earth provided they are built solid without wood reinforcement.

Earth walls for dwelling rooms and for stables should be thick enough to provide sufficient heat insulation for the particular climatic conditions.

FLOORS

Solid floors in place of the usual wood floors, and beams of steel or concrete, should normally be supported only on brick, stone, or concrete walls; that is, on basement walls, or on supports which are equally solid.

ROOFS

Pitched roofs may be either gable ended, hipped, or half hipped. Attic walls may be carried up to a height of 1 metre.

Roofs should project sufficiently at eaves and gable ends, in order to give thorough protection to the walls from rain.

The roof covering should normally be of tiles. With detached and semi-detached buildings clay shingles, as well as straw- or reed-thatch, may be used, provided the regulations governing the latter coverings are adhered to.

COB, PISÉ, AND STABILIZED EARTH

Clay shingles should be made from clay and strong rye straw, and should be at least 20 cm. thick; nowhere should the clay layer be less than 2 cm. thick on the underside of the roof. The roof covering can be made secure by stretched wires, netting, willow branches, and so forth. The ridge may be covered by placing rolls of straw and mud across it at right angles, or by purpose-made ridge tiles, or in any other equally efficient way. The junction of the walls with the underside of the roof should be properly sealed.

CHIMNEYS

Earth should not be used in the construction of chimneys. Mud walls and the masonry for chimneys should be bonded by vertical rebates and not by toothing.

RENDERINGS

Renderings should not be applied until the earth is sufficiently dried to obviate further settlement and shrinkage cracks. Cob walls should not be rendered until at least a year after completion. These and other earth walls which cannot be rendered for some time should be protected with a waterproof coating, such as a limewash with skimmed milk.

Renderings on external walls of habitable rooms should normally be in two coats and be waterproof. On no account should the finishing coat be of a stronger mix than the undercoat. In order to achieve a good key for the rendering, the surfaces of the walls and of the undercoat should be scored and indented so that the rendering may obtain a hold on the exposed fibrous or stony aggregate, and in the indentations. External renderings may be omitted on buildings of lesser importance if the walls are protected with a wash.

Special internal plastering is unnecessary if the walls are even and are provided with a thin layer of mud plaster. Walls of stables and of other rooms whose use renders them liable to damage and wetting should be protected internally by wood lining or other similar means.

DOORS AND WINDOWS

Lintols for doors and windows spanning not more than 1·20 metres may be of wood or concrete, provided that they have a

WALL SECTION CONFORMING
TO POSEN RECOMMENDATIONS

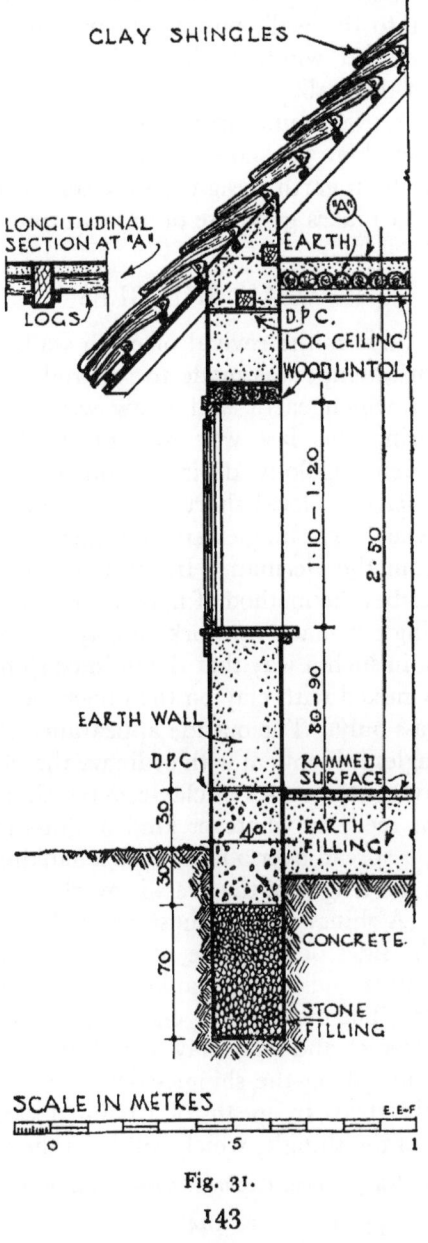

CLAY SHINGLES

LONGITUDINAL
SECTION AT "A"

LOGS

EARTH

D.P.C.
LOG CEILING
WOOD LINTOL

1·10 – 1·20

2·50

·80 – ·90

EARTH WALL

D.P.C.

RAMMED
SURFACE

·30

·30

EARTH
FILLING

·70

CONCRETE.

STONE
FILLING

SCALE IN MÉTRES

E. E=F

0   ·5   1

Fig. 31.

143

bearing on the walls of at least 25 cm. to distribute the pressure equally. For greater spans the pressure should be taken by breastings or masonry piers. Door and window frames should not be built into the wall, but should be inserted into rebates. External doors and windows should be placed flush with the outside face of the wall.

External windows should open outwards.

The drawing (Fig. 31) shows a typical cross-section through an external wall in earth designed to accord with these recommendations. It makes good use of unsawn timbers.

CLAY SHINGLES

At the same time as the revival of earth walling after the first world war, an attempt was made to reintroduce an old method of roofing, in which earth and straw were used to form clay shingles. During this last war Mr. Ernst May, an architect practising in Kenya Colony, devised a similar method of roofing, in order to overcome local shortage of material (Fig. 32), and, as we have seen, clay shingles are also mentioned in the above translation from the German. In an article from *Country Life* [1] Mr. May describes the method of making clay shingles as follows:

"A clay-shingle is made by working clay into a layer of straw, reed, or grass, in such a way that the finished shingle consists of a layer of straw mixed with clay on the underside, while the top is formed by grass only. The outside appearance of a roof covered with clay-shingles is identical with ordinary thatch, but the inside shows no straw, only courses of clay-covered shingles.

"The straw, reed, or whatever kind of grass may be used for their making, is spread over a table with a flanking board on each side. Half of the length of the grass is overlapping the near side of the table. A shingle-stick, whose pointed ends project a few inches over the sides of the table, is now inserted into the slots cut out of the near ends of the flanking boards and well bedded in clay-mortar. The overlapping ends of the grass are bent back over the shingle-stick and more clay is applied to a strip approximately 14 in. wide along the shingle-stick. This clay is carefully worked into the straw by means of a pointed stick and then the upper surface of the shingle, which will form the underside of the

[1] Ernst May, "Clay Shingle Roofing," *Country Life*, January 5th, 1945.

Pl. 48.—The front view of the Children's Holiday Camp, Nairobi.

Pl. 49.—The garden view of the Children's Holiday Camp.

Pl. 50.—Reeds are spread on a table so that half their length overlaps the near end. The shingle-stick is also seen in position.

Pl. 51.—The overlapping ends of the reeds are bent back over the shingle-stick.

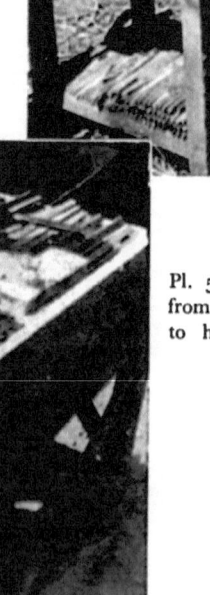

Pl. 52.—Two hanging pegs project from the under side of the shingle to hold the shingle on the roof battens.

Pl. 53.—Finished shingles stacked.

Pl. 54.—Clay shingles hung as ordinary tiles, showing the clay finish on the under side to the roof.

Pl. 55.—Shingles in position.

Pl. 56.—Building in door frames.

Pl. 57.—The "Umbrella" roof construction out of contact with the walls and forming a convenient verandah.

Pl. 58.—The finished hut.

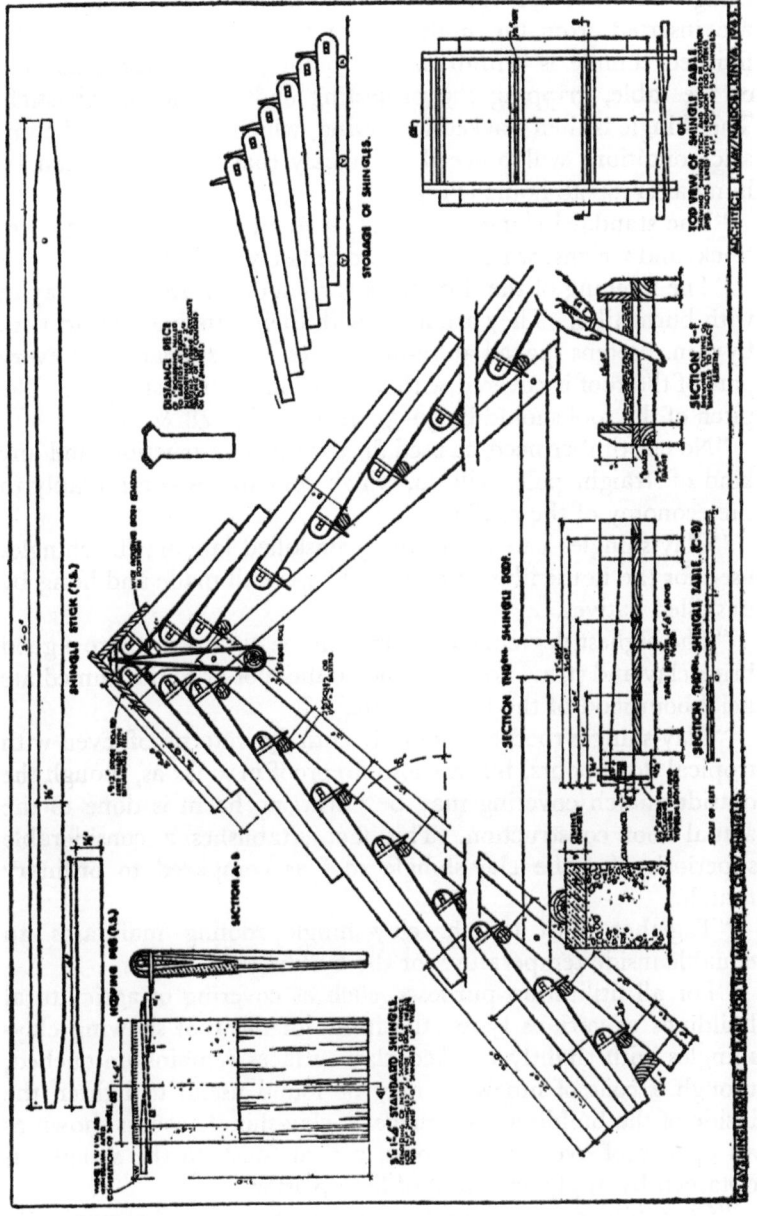

Fig. 32.

shingle on the roof, is smoothed with a trowel. Two pointed pegs are inserted from the underside of the shingle, and then the finished shingle is withdrawn by pulling it over the near side of the table, gripping the projecting ends of the shingle-stick. The shingle is then stacked for drying, which under normal climatic conditions will proceed so quickly that the shingle can be hung after 24–48 hours (Pl. 53).

"The standard shingle measures 3 ft. by 1 ft. 6 in., is 3–4 in. thick, and weighs, when dry, approximately 18 lb.

"The hanging of the shingles is carried out in the same way as with burnt tiles. They are hung with their hanging-pegs to 1-in. by 2-in. battens spaced at 12-in. centres. This means that every part of the roof is covered with a threefold layer of shingles. The pitch of the roof should be 50–55 degrees (Pls. 48 & 49).

"No cut timber need be used for the roof construction, and any kind of straight poles will do, which contributes considerably to the economy of the roofing.

"Clay-shingles can be made by unskilled labour; the shingles used for the house illustrated (Pl. 48) were all made and hung by unskilled natives.

"The highest degree of economy will be obtained where a good brick clay and straw are available on the spot or in the immediate neighbourhood of the building site.

"Clay-shingle roofs are not only entirely waterproof, even with tropical downpours, but are also fireproof in so far as, though the outside thatch covering may be burnt, no harm is done to the actual roof construction. This fact establishes a considerable superiority for the clay-shingle roof as compared to ordinary thatch.

"Together with thatch, clay-shingle roofing maintains an equable inside temperature for the buildings so roofed.

"For all utilitarian purposes, such as covering of agricultural buildings of various types, the inside of the roof showing clay-shingles with slightly cracked clay surfaces remains untouched, though a coat of limewash may be found useful to lighten the inside of the building. Whenever a clay-shingle roof is shown as an open roof over living-rooms, a neat finish to the shingles is obtained by applying a coat of lime plaster.

"Clay-shingle roofs, like thatched roofs, merge harmoniously into the landscape."

SOME PRACTICES ABROAD

Apart from the use of earth-building methods in times of emergency, it is clear that in many parts of the world walls of Pisé or of sun-dried earth blocks have been, and will continue to be, accepted methods of building. Improvements will undoubtedly be made by the control of earth mixes, with or without stabilizing agents; and by the introduction of efficient damp-proof courses, ant courses, and protective coverings.

This is certainly true of a number of the Colonies, where primitive earth-building techniques have been adopted as a basis for some research.

THE COLONIES

In the West Indies, for instance, Pisé, and "Tapia," which is a refinement of wattle and daub, are two amongst other traditional building methods which are used today. In British West Africa walls are commonly built of Pisé and sun-dried bricks, and of kneaded lumps of a lateritic earth, known locally as "Swish," compressed course upon course, as in Cob walling. In parts of British East Africa similar techniques, including sun-dried bricks and wattle and daub, are used.

While it is obvious that primitive hutments of earth do not constitute a satisfactory standard for native housing, it is also obvious that many of the improvements which must be effected concern elements of design and construction other than the walling material itself; houses with earth walls have the advantage that if there is also a suitable roof, indoor temperatures remain equable, even in extremes of heat and cold; also, with earth walling, full use can be made of local materials.

Experiments in stabilizing soils, in the design of simple mechanical equipment such as machines for making earth blocks, and in the possible use of "D.D.T." and other insecticides, may enable earth building to make a real contribution to the improvement of housing in the Colonies.

UGANDA

The Pisé building illustrated (Pl. 58) is a model hut designed as an example for native practice in Uganda,[1] and the illustration

[1] G. Gillanders, "Rural Hygiene in the Tropics," *Journal of the Royal Sanitary Institute*, Vol. LX, No. 6, December 1939.

also shows a similar house in course of erection. The roof is supported on posts and forms an umbrella to the free-standing Pisé walls, protecting them from sun and rain. A ventilation space is provided as shown, between wall head and rafters, giving access to ant runs along the tops of walls and also dissociating the timber roof members from the walls. It is known that within three weeks the natives can be taught to erect substantial buildings in Pisé with simple shuttering. The design for the shuttering, and information on methods of building in Pisé, are the subject of official publications in Uganda by the Uganda Medical Department.

Wattle and daub rather than Pisé is, however, traditional to Uganda, and model huts of this kind have also been built to show that it is possible to overcome the harbouring of spirillum ticks, bugs, and even rats, for which they have an unenviable reputation. By sealing cracks and by building the walls reasonably straight, with a plastering of swamp sand and cow dung, subsequently limewashed, it has been proved that wattle and daub huts can be built and maintained vermin-proof.

The floors are often made of well-rammed, crushed lateritic earth, smeared regularly with cow dung, or of white ants' nests broken up, watered, and well rammed. Bitumen cutbacks and bituminous emulsions laid on a thin layer of fine gravel have also been used successfully for flooring, whilst if a layer of these materials be laid on the walls at floor level it·can be sealed to the bitumen gravel floor to provide a continuous damp-proof membrane over the whole floor area. This damp-proof layer is said to provide also some measure of protection against white ants. Where it is available, however, metal forms the most complete barrier to white ants and can be used both at damp-proof course level and under the wall plates on top of the wall. If the metal layer projects from the wall and is turned down, the ant is prevented from forming its characteristic tunnel of earth, by means of which it can reach the wall above any normal damp-proof course or the timber wall plate, or roof members. Wherever possible, contact between earth and wood should be avoided, and outside steps and other features by which the ant could reach the walls should be separated from the main structure by a narrow air space.

SOME PRACTICES ABROAD

NIGERIA [1]

In Nigeria the traditional use of earth for building is not confined to small houses only, for in the Haussa district, and especially in Kano, large buildings such as those illustrated in Pls. 59 and 60 are built in earth. The character of the buildings, and especially of the vaulted arches, is reminiscent of Saracenic masonry construction, and does, in fact, derive from the Mahommedan element in the people of the district.

The older and more skilled craftsmen in Nigeria do not travel far, and instruct only their own children, with the result that after a few generations the methods differ slightly from city to city. Europeans have difficulty in influencing the methods of the native craftsman, who is suspicious of innovators, who have not an intimate knowledge of his craft, and who believes that anyone not born to the trade cannot possibly know enough to criticize.

The earth for building varies, but the local head builders usually know where the best earth is to be found. Native bricks (called "Tubali" in some districts) are made of this earth, and are a form of sun-dried brick. The shape and size varies from province to province, but is usually in the form of a rough cone. The earth is first wetted and trampled until it reaches a workable consistency; it is then left for one or two days, when it is again wetted and trampled, and finally formed without moulds into Tubali, which are allowed to dry for about ten days or more. They are laid with the larger end facing alternately inward and outward. A clay covering is applied to the wall as the work proceeds.

Mortar is made by wetting and trampling the earth once, and covering it with a layer of horse manure; after three days, during which the earth is kept moist, the manure is thoroughly mixed with the earth by further trampling. The process is repeated four times during a period of two to three weeks. If manure is not available, short lengths of grass, fresh or dried, cut into 3-in. lengths, are used.

Traditionally, walls of mud were built thicker than was structurally necessary. No doubt this was from a desire for safety and for additional heat insulation, and also possibly because slave

[1] A. F. Daldy, A.M.I.C.E., "Temporary Buildings in Nigeria," P.W.D. Technical Paper No. 10.

labour was so cheap. Very often, and especially in the vaulted construction of the Haussa district, a form of reinforcement was provided by jungle sticks or palm fronds. The wood of several hardwood trees in Northern Nigeria, being not readily susceptible to termite attack, is used in building on that account.

Internal plastering is usually in two coats, in addition to the covering of the Tubali, which is applied as the work proceeds. The first plaster coat is of ordinary mud mortar; the second coat is of red earth and sand. This second coat can be finished with a trowel, and provides a satisfactory base for whitewash, if time is first allowed for drying out. For ceilings a special plaster of black earth and grass is used which adheres better to the laths used in the ceiling construction. When this coat, which is put on wet, has dried and finished cracking, a second coat of red earth and sand is added.

For external plastering one coat of mud mortar is used, followed by a final coat of a waterproofing compound. In important buildings, such as the palaces of the Emirs, in the Haussa district, soffits of arches are sometimes treated with special plasters. One such plaster contains small fragments of quartz set on gum arabic and has an iridescent appearance; another includes mica particles. Again, plates or bowls are set in the plaster, and are put there, it is said, not only for decoration, but also as an indication of stability, for when the building becomes unstable or settles, the bowls fall off. This surface decoration, like the vaulting, is characteristic of Saracenic architecture.

The necessity for protecting earth walls against rain led the native to discover for himself a number of waterproofers and special finishes. After experimenting with these and other protective coverings, it was found that Dutch Plaster suited best the particular conditions. This plaster is composed of lateritic earth, sand, mud, and cement, in the typical proportions, 5 lateritic earth : 3 sand : 2 mud : 1 cement, or 4 sand : 6 mud : 1 cement, depending on the cohesion of the mud and the availability of the lateritic earth.

GOLD COAST

On the Gold Coast the villagers use the lateritic soil, known locally as swish, to make lumps about 6 to 8 in. in diameter, which are thrown down upon each other and pounded to form a course about 12 in. wide by 12 in. high. The technique is, of course, similar to Cob, and, as with Cob, the walls have to be protected during construction from rain and, in the tropics, from the sun; this is often done with leaves.

Door and window frames are fitted in after the walls are complete, openings having been left during construction. Inside, the walls are plastered with Kaolin clay. If the owner has enough money, he uses a cement and sand rendering externally; if he has not, he uses clay. The clay, of course, usually adheres fairly well, while the cement rendering shrinks, becomes hollow, and finally falls off.

Some experiments in controlling mixes, and in stabilizing swish, have been made in order to overcome the excessive cracking which results from native methods, and to provide a more durable wall. Blocks 18 in. by 9 in. by 6 in. of 20 to 1 swish and cement have been found to make satisfactory walls only 6 in. thick. For external treatments limewash, cement wash, and tar are being tried; internally, the walls may be plastered. In using a Stabilized Earth and reducing the thicknesses of walls, it should not be forgotten that essential features of traditional earth building are the heat insulation provided by thick walls, and the equable inside temperature which results from their great mass and high heat capacity. Walls must either remain thick or heat insulation of some other kind must be introduced. Similarly, wide overhanging eaves and verandahs which give protection against the sun should not be omitted. Not only do they give character: they have also their essential functions in tropical climates.

RUSSIA

In Russia, in the Ukraine, and in the Ukrainian settlements in Canada, some of the most beautiful of earth buildings are to be found. In these places the craft is a living tradition; there is no question of earth houses being in any way inferior, indeed those who have lived in both earth and timber houses in Canada speak favourably of the former. Of course, it is especially necessary in these climates to ensure that the walls are built on properly drained ground and to provide adequate protection at the eaves and at the base of the wall.

Some of the methods of construction with earth in combination with timber and with earth alone are described below in extracts translated from *Dwellings of the People in the Ukraine*, by P. G. Yurchenko.[1]

"The architecture of the people in country districts employs local structural materials exclusively, and in that respect is an instructive example of the variety of uses to which clay, wood, straw, reeds and other local structural resources can be put. In the recent past, in the construction even of the more complicated village buildings it was customary to dispense with imported materials, with the exception of glass. Frequently, even iron nails were replaced by wooden pins, straw and osier. At the same time, the method of construction was distinguished by extreme simplicity. The architect of the people had at disposal very restricted means for the solution of structural problems, and he had, therefore, to show great inventiveness and a knowledge of every type of material employed.

"In the Ukraine it has for long been customary to build two types of dwelling, namely log huts and earth huts. It would be difficult to say which of these methods is to be considered the more typical of Ukrainian vernacular architecture. Both existed in olden times and both are equally characteristic of different parts of the Ukraine. The architect of the people uses wood in preference to other materials because of its comparative lightness and low thermal conductivity. Communities who live on the edge of a forest always build in wood only; but in the conditions of the steppes the only suitable substitute for wood is earth alone or combined with straw and wood. Thanks to the existence of clearly marked forest and steppe zones in the Ukraine, the two methods of construction are equally widely used there.

"At a comparatively late date, the architects of the Ukraine began to introduce widely a more complicated type of construction, namely, framed houses. This type was resorted to owing, chiefly, to the need for economizing wood.

"Three types of log house may be mentioned which are typical of the different forest and forest-steppe regions of the Ukraine.

"In the forest region (Polessie) they make the huts of roughly-squared full beams or beams sawn in two, and in both cases coat

[1] P. G. Yurchenko, *Dwellings of the People in the Ukraine.* 1941. Extracts translated by G. N. Gibson. Reproduced in the *Journal of the R.I.B.A.*, Vol. 52, 3rd series, No. 10, August 1945.

the walls inside and outside with clay. The most northerly districts are an exception; there they use clay simply to fill the chinks between the logs and the latter are left in their natural state or are whitewashed (Fig. 33).

"In the forest-steppe region they build the wooden walls of squared beams, 12 to 18 cm. deep, and 20 to 25 cm. high, and apply the clay coating to both sides (Fig. 34).

"The third type, departing from the usual practice, and found sometimes in certain parts of the Carpathians, is a structure of neat squared beams of oak, alder or lime of the same dimensions,

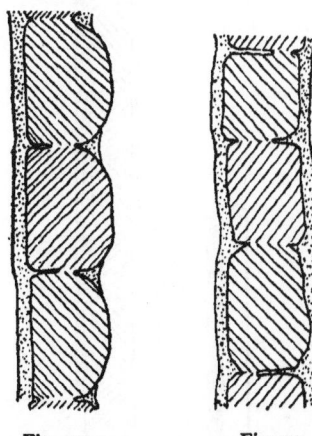

Fig. 33.   Fig. 34.

Fig. 33.—Wall of logs with clay filling and coating (Polessie forest district).
Fig. 34.—Logs with clay coating (districts of Kiev, Poltava, Podolia), used in forest-steppe region.

for example, as the preceding type. In this case the interior wall surfaces of the living quarters are finished off with particular care. The beams are fitted very accurately and the inside wall surface is not coated with clay. Such walls are easily kept clean with soap and water.

"Foundations in the customary sense, with footings deep in the ground below the freezing line, are not usually made in wooden buildings of the Ukraine. The first logs are laid on blocks of wood, tree stumps or large stones placed at a slight depth in the ground to increase stability. They are sometimes even placed simply on a bedding consisting of a single layer of stones laid on the site when it has been made level. A compact foundation of stones under the house is rarely used. With foundations constructed in this way, there is danger of freezing of the wall and of the part of the floor near the wall. For the sake of warmth, the lower part of the wall is increased in thickness as shown in Fig. 35, i.e. by a projecting plinth (or prisba). The plinth is made of earth rammed between the wall of the hut and a low fence of planks or brushwood parallel to the wall. The plinth as well as the wall of the house is coated with clay.

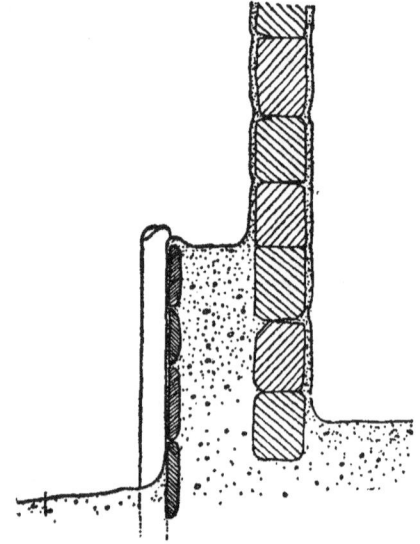

Fig. 35.—Construction of a projecting plinth or "prisba" (Poltava).

"The scarcity of wood, especially in the steppes, has for long made it necessary for the village architect to seek more rational and economic methods in the use of woods. One such method is to construct the hut as a frame with wall panels of other materials. This method of construction has been adopted very extensively in the Ukraine.

"Wooden posts are set in holes in the ground, 1 to 1·5 m. apart, and at the top they support a frame formed of two to four beams (Fig. 36). The space between the posts is filled with spare timbers, the ends of which fit into grooves made in the posts or formed by two battens nailed to the posts. A batten is nailed to

Pl. 59.—The changeless streets of Old Kano, Nigeria. Little girls selling cakes in a street running from the camera.

Pl. 60.—City houses in Kano, Nigeria.

the timber panel or it is notched with an axe or a short wooden wedge is driven in. After having been prepared in this way, the wall is coated with a layer of earth well mixed with straw (Fig.

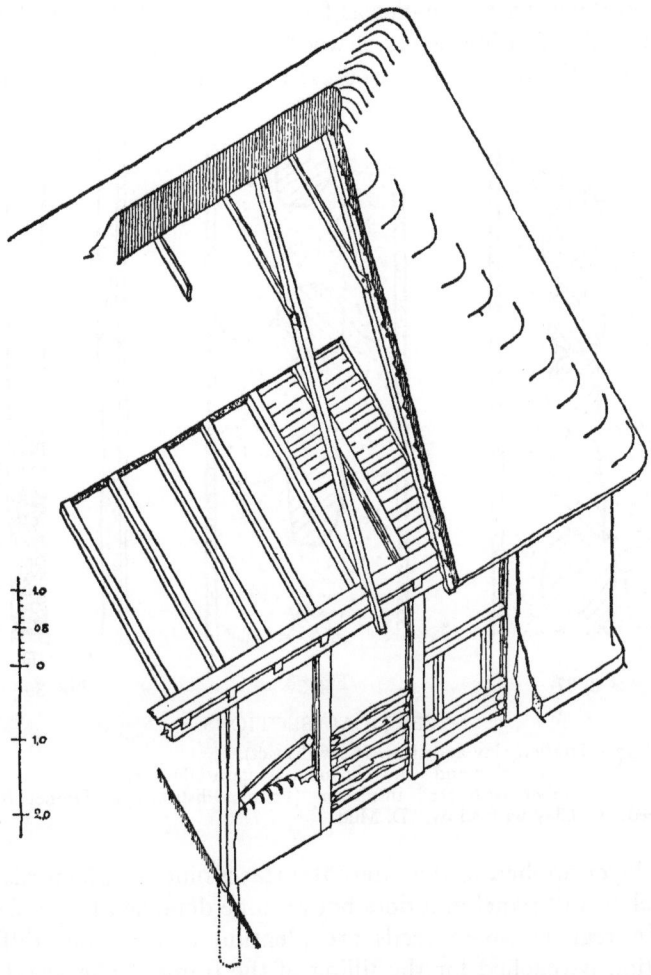

Fig. 36.—Frame construction of a peasant's cottage (village in Kiev district).

37). This method of 'filling' between the posts requires, however, a comparatively large quantity of wood, and, in addition, the clay coating applied to the compact surface loosens when the wood dries, and frequently falls off. The defects of the compact wall surface between the posts are absent from the following type of construction. The first board of the wall panel is laid on the ground and on this is laid a layer of lumps of clay mixed with plenty of straw. On this another log of wood is laid and hammered down with a mallet as far as it will go into the clods beneath. Alternating layers of wood and clay are obtained in this way and the quantity of wood is reduced to half. A neat, thin

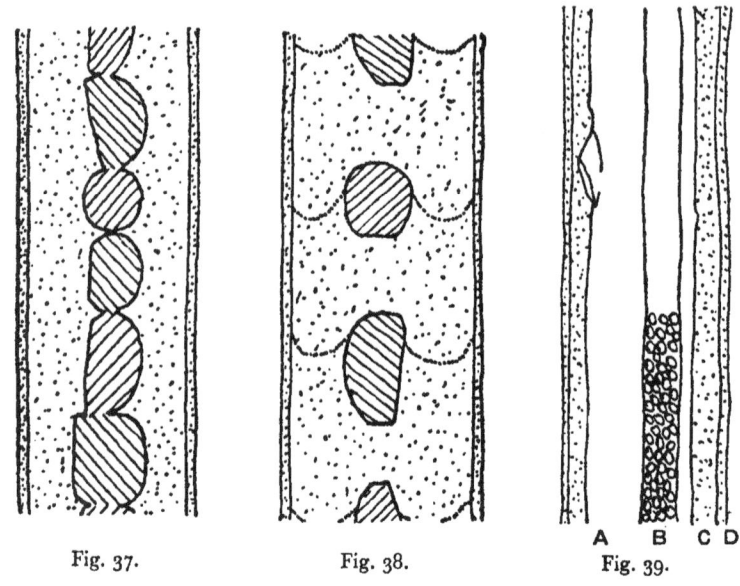

Fig. 37.   Fig. 38.   Fig. 39.

VERTICAL SECTIONS.

Fig. 37.—Timber, clay and straw (Kiev district).
Fig. 38.—Timber, clay and straw (Podolia, Western Ukraine).
Fig. 39.—Timber with reed insulation (Poltava district). A. Timber frame. B. Reed. C. Clay with straw. D. Mud.

clay layer applied to the external surfaces binds satisfactorily with the clay wall panel and does not become detached (Fig. 38).

"In regions where reeds are plentiful a somewhat different solution is reached for the filling of the frame. The wood used for the panel is of small section, i.e. about 3 to 4 in. in diameter. For heat insulation of the wall a vertical layer of reeds, about 2 in. thick, is placed against the wood at the outer wall surface and pressed against it by thin laths; the clay covering is applied to the reeds (Fig. 39).

"Most often, however, in those regions, the walling between the posts is constructed entirely of clay. Between the posts three holes about 2 in. in section are placed horizontally in prepared openings; to these bundles of reeds are secured with cord in a vertical position and with a staggered arrangement. Clay mixed with chopped straw is pressed into the gaps between the bundles of

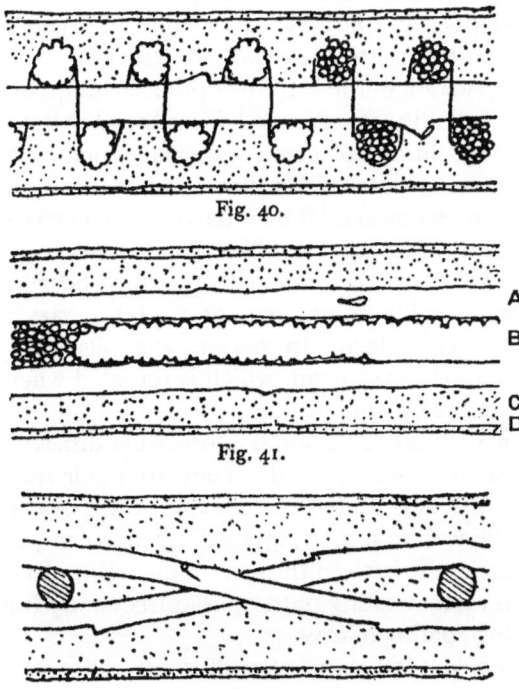

Fig. 40.

A
B
C
D

Fig. 41.

Fig. 42.

HORIZONTAL SECTIONS.

Fig. 40.—Reeds and clay (Poltava district).
Fig. 41.—Reeds and clay (Poltava district). A. Poles. B. Reed. C. Clay with straw. D. Outside earth coating.
Fig. 42.—Plaited wattles and clay (district of Dnyepropetrovsk).

reeds on both sides and the wall surface is then smoothly finished with a coating of clay (Fig. 40).

"Comparatively rarely the space between the posts of the frame is filled with reeds placed vertically and compressed between two poles, and this is coated on both sides with a layer of clay (Fig. 41).

COB, PISÉ, AND STABILIZED EARTH

"The types of wall construction which have been considered are those of the heated living accommodation; the walls of the unheated part, however, in most cases are without a clay coating. The walls in such cases, whether the hut be of the frame type or not, are for reasons of economy left exposed. Sometimes the walls of the unheated rooms are simply of wattle between posts and are coated with clay on both sides (Fig. 42).

"In the steppes and forest-steppe region the type most widely extended is one consisting of wall construction of earth alone. For the construction of the wall they prepare lumps of clay mixed with a large quantity of straw and these they lay flat in the wall. They make the lumps of earth somewhat longer than the required thickness of the wall, so that when the wall has dried out the surface can be finished off by removing any marked irregularities formed during laying. As they lay the lumps of earth in the wall they press down the ends with pieces of board in order to close up any possible chinks and as a preliminary measure of obtaining an even surface. In this way the rolls of clay are given an almost semi-circular form, which is revealed when old walls are demolished (Figs. 43, 44).

"In Podolia (Western Ukraine) a somewhat different method of laying the earth 'blocks' is used. They are made from mud of a very stiff consistency, and are of a more or less oval shape; they are laid in the wall in a 'herringbone' arrangement, i.e. with one course sloping to the left and the next to the right. In this way the wall is given an interesting pattern, and frequently it is left without a finishing coating of clay.

"The rammed earth type of wall construction is less widely used. The earth is rammed in a semi-dry state in vertically movable forms, and in some of the layers in the interior of the wall wood is placed as reinforcement.

"In regions where formerly there were military settlements, a widespread method of wall construction made use of unfired earth blocks. Similar blocks, but of a large type, measuring about 11 in. by 11 in. by 5 in., were made and were dried before being built into the wall. This method was evidently introduced from outside, but it was not widely used, since it necessitated a considerable amount of work and could not be undertaken by the peasant builder alone. With frame and clay wall construction it is essential to construct a plinth, as not only must the lower part of the wall be protected from freezing, but also the penetration of damp through the clay wall to the interior must be prevented. For these reasons and also to increase the durability of the clay wall, its thickness is not uniform; at the lower part the wall thickness is from about 3¼ to 4 ft., while at the top it is about 2 ft. to

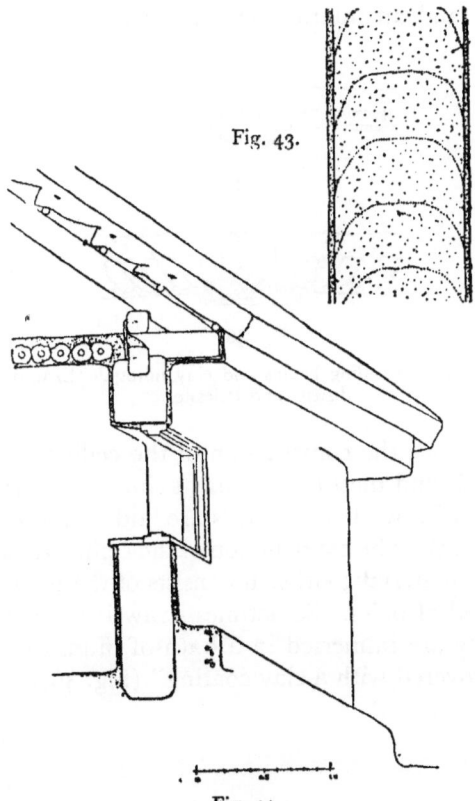

Fig. 43.

Fig. 44.

Figs. 43, 44.—The construction of a peasant's cottage walls from earth lumps (Kiev district).

1⅜ ft. The construction of clay walls is very laborious, and much time has to be allowed for drying and shrinkage of the wall; the latter is considerable, usually about 2 in. in about 3 ft. 3 in. of height."

The normal roof covering for these houses is straw or reed thatch, but sometimes wood shingles are used. The early nineteenth-century practice of using clay on boards laid on an elaborate construction of wood beams has given way to lighter forms of roofing, but clay is still used for the plastering of the ceilings. Of these the author says:

"The span between the external walls of the hut or between an external wall and an interior bearing wall is rarely more than

Fig. 45.

Fig. 46.

Figs. 45, 46.—Sections of ceiling beams and clay fillings. (Examples from Kiev district and Polessie.)

about 16 ft.; hence the construction of the ceiling has gradually been simplified, and most often consists simply of joists spaced at about 3 ft. or 4 ft., while the boards are laid on battens nailed to the joists (Fig. 45). The construction of the ceiling depends on the local available materials; either it consists of the usual boards, or else it is formed of poles. Sometimes straw is twisted round the poles, and they are immersed in a bath of mud, then placed in position and covered with a clay coating" (Fig. 46).

Index

161

INDEX

INDEX

INDEX